Wheat Production in Changing Environments

Wheat Production in Changing Environments

Edited by:

Akansha Singh

Delve
PUBLISHING

www.delvepublishing.com

Wheat Production in Changing Environments

Akansha Singh

Delve Publishing
224 Shoreacres Road
Burlington, ON L7L 2H2
Canada
www.delvepublishing.com
Email: orders@arclereducation.com

ISBN: 978-1-77469-127-4 (Hardcover)

ABOUT THE EDITOR

Dr. Akansha Singh, is presently working as Associate Professor & Head, Department of Genetics and Plant Breeding, College of Agriculture, Parul University, India. She graduated from the Banaras Hindu University, with doctorate in Genetics and Plant breeding in the year 2012. She has also worked as Research Scientist in Bioved Research Society, Allahabad, India. She was working as Postdoctoral fellow in University of Mumbai before joining her current position. She has authored a number of national and international publications.

TABLE OF CONTENTS

LIST OF FIGURES

LIST OF TABLES

LIST OF ABBREVIATIONS

APX	Ascorbate Peroxidase
ATSDR	Agency for Toxic Substances and Disease Registry
AYTs	Advanced Yield Trials
BNI	Biological Nitrification Inhibition
CC	Climate Change
CCC	Canadian Climate Centre
CEC	Cation Exchange Capacity
CIMMYT	International Maize and Wheat Improvement Centre
DNA	Deoxyribonucleic Acid
EEC	European Economic Community
FAO	Food and Agriculture Organization
GCM	Global Climate Models
GFHI	Geophysical Dynamics Laboratory High Scenario
GHGs	Greenhouse Gases
GHI	Global Hunger Index
HMF	Hydroxy Methyl Furfural
ILK	Indigenous and Local Knowledge
INDC	Intended Nationally Determined Contribution
IPCC	Intergovernmental Panel on Climate Change
IUPAC	International Union of Pure and Applied Chemistry
IWIN	International Wheat Improvement Network
JECFA	Joint Expert Committee of Food Additives
NDVI	Normalized Difference Vegetation Index
NPP	Net Primary Production
OECD	Organization for Economic Co-operation and Development
PBI	Plant Breeding Institute
PPT/PET	Potential Evapotranspiration Rate
PSII	Photosystem II

R&D	Research and Development
ROS	Reactive Oxygen Species
SCOPE	Scientific Committee on Problems of the Environment
SHW	Synthetic Hexaploid Wheat
SOC	Soil Organic Carbon
SOD	Superoxide Dismutase
SOM	Soil Organic Matter
SSPs	Socio-Economic Pathways
SYN-DER	Synthetic Derived
UAV	Unmanned Aerial Vehicle
USDA	U.S. Department of Agriculture
WSBM	Wheat Soil-Borne Mosaic
WSM	Wheat Streak Mosaic

GLOSSARY

A

Accelerators – a person or thing that causes something to happen or develop more quickly.

Agroforestry – Agroforestry is a land use management system in which trees or shrubs are grown around or among crops or pastureland.

Agronomy – is the science and technology of producing and using plants in agriculture for food, fuel, fiber, recreation, and land restoration.

Antioxidants – antioxidants are compounds that inhibit oxidation, a chemical reaction that can produce free radicals and chain reactions that may damage the cells of organisms.

Assimilate – take in and understand fully (information or ideas).

B

Bioclimatic – relating to the interrelation of climate and the activities and distribution of living organisms.

Biotechnology – It is a technology that uses biological systems, living organisms, or parts of this to develop or create different products.

C

Continent – is one of several very large landmasses. Generally identified by convention rather than any strict criteria, up to seven regions are commonly regarded as continents.

D

Deforestation – Deforestation, clearance, clearcutting, or clearing is the removal of a forest or stand of trees from land that is then converted to nonforest use.

Desertification – is a type of land degradation in drylands involving loss of biological productivity caused by natural processes or induced by human activities.

Drought – a drought is an event of prolonged shortages in the water supply, whether atmospheric, surface water or groundwater.

E

Ecosystems – An ecosystem is a community of living organisms in conjunction with the nonliving components of their environment, interacting as a system.

Ecumenism – also spelled oecumenism, is the concept and principle in which Christians who belong to different Christian denominations work together to develop closer relationships among their churches and promote Christian unity.

Endosperm – is a tissue produced inside the seeds of most of the flowering plants following double fertilization.
Evapotranspiration – is the sum of evaporation and plant transpiration from the Earth's land and ocean surface to the atmosphere.

F

Fertilization – also known as generative fertilization, insemination, pollination, fecundation, syngamy and impregnation, is the fusion of gametes to initiate the development of a new individual organism or offspring.
Fluxes – is a chemical cleaning agent, flowing agent, or purifying agent.
Fungicides – are biocidal chemical compounds or biological organisms used to kill parasitic fungi or their spores.

G

Generic – characteristic of or relating to a class or group of things; not specific.
Genetic Diversity – Genetic diversity is the total number of genetic characteristics in the genetic makeup of a species, it ranges widely from the number of species to differences within species and can be attributed to the span of survival for a species.
Genome – in the fields of molecular biology and genetics, a genome is all genetic material of an organism
Genotypes – the genetic constitution of an individual organism.
Grassland – areas where the vegetation is dominated by grasses; however, sedge and rush families can also be found along with variable proportions of legumes, like clover, and other herbs.

H

Herbicides – also commonly known as weedkillers, are substances used to control unwanted plants.
Herbivores – an animal that feeds on plants.

I

Imitates – take or follow as a model.
Inoculum – a substance used for inoculation.

L

Landraces – a pig of a white lop-eared breed, originally developed in Denmark.
Livestock – is commonly defined as domesticated animals raised in an agricultural setting to produce labor and commodities such as meat, eggs, milk, fur, leather, and wool.

M

Mediterranean Sea – is a sea connected to the Atlantic Ocean, surrounded by the Mediterranean Basin and almost completely enclosed by land.
Metabolism – is the set of life-sustaining chemical reactions in organisms.

Microorganisms – Microorganisms or microbes are microscopic organisms that exist as unicellular, multicellular, or cell clusters.
Mildew – a thin whitish coating consisting of minute fungal hyphae, growing on plants or damp organic material such as paper.

N
Nitrogen – a nonmetallic element that makes up about 78% of the atmosphere by volume, occurring as a colorless, odorless gas.
Nutrient – a nutrient is a substance used by an organism to survive, grow, and reproduce.

O
Oddity – a strange or peculiar person or thing.
Osmolytes – osmolytes are low-molecular-weight organic compounds that influence the properties of biological fluids.

P
Painstakingly – with great care and thoroughness.
Pathogens – a bacterium, virus, or other microorganism that can cause disease.
Pertinent – closely connected with the subject being discussed.
Pesticide – They are substances that are meant to control pests. The term pesticide includes all of the following: herbicide, insecticides nematicide, molluscicide, piscicide, avicide, rodenticide, bactericide, insect repellent, animal repellent, antimicrobial, and fungicide.
Phenology – is the study of periodic events in biological life cycles and how these are influenced by seasonal and interannual variations in climate, as well as habitat factors.
Phenology – is the study of periodic plant and animal life cycle events and how these are influenced by seasonal and interannual variations in climate, as well as habitat factors.
Phenotypic – relating to the observable characteristics of an individual resulting from the interaction of its genotype with the environment.
Phosphorus – phosphorus is a chemical element with the symbol P and atomic number 15.
Prairie – are ecosystems considered part of the temperate grasslands, savannas, and shrublands biome by ecologists, based on similar temperate climates, moderate rainfall, and a composition of grasses, herbs, and shrubs, rather than trees, as the dominant vegetation type.
Precipitation – is any product of the condensation of atmospheric water vapor that falls under gravity from clouds.
Progeny – a descendant of the descendants of a person, animal, or plant; offspring.
Riboflavin – also known as vitamin B_2, is a vitamin found in food and used as a dietary supplement.

S
Sappers – a soldier responsible for tasks such as building and repairing roads and bridges, laying and clearing mines, etc.

Savanna – is a mixed woodland-grassland ecosystem characterized by the trees being sufficiently widely spaced so that the canopy does not close.
Senescence – the condition or process of deterioration with age.
Sketched – make a rough drawing of.
Soluble – (of a substance) able to be dissolved, especially in water.
Sterility – the quality or condition of being sterile.
Superfluous – unnecessary, especially through being more than enough.
Sustainability – Sustainability is the capacity to endure in a relatively ongoing way across various domains of life.

T

Temperate – the temperate or tepid climates of Earth occur in the middle latitudes, which span between the tropics and the polar regions of Earth.
Thiamin – thiamine, also known as thiamin or vitamin B_1, is a vitamin found in food and manufactured as a dietary supplement and medication.
Tillage – the preparation of land for growing crops.
Tillage – tillage is the agricultural preparation of soil by mechanical agitation of various types, such as digging, stirring, and overturning.
Topographical – is the study of the forms and features of land surfaces.

V

Vicissitudes – a change of circumstances or fortune, typically one that is unwelcome or unpleasan

PREFACE

This book takes the readers through several different stages of the production of wheat. This book sheds light on the several aspects of global wheat production, impact of climate change on grassland, environment and crop production, progress and challenges of wheat production, climate change and food security, breeding strategies and quality enhancement under climate change, and crop adaptation to climate change.

The first chapter stresses on the basic overview of the production of wheat so that the readers are clear about the philosophies behind that form the utmost basics in the field. This chapter will also emphasize the history of wheat, plant breeding, types of wheat, traditional strategy for wheat production, ecology and climatic requirements.

The second chapter takes the readers through the concepts of global wheat production. This chapter will provide highlights on the effect of phosphorus on root signaling of wheat under different water regimes, wheat sensitivity to nitrogen supply under different climatic conditions, challenges to safe wheat storage.

Then, the third chapter explains the impact of climate change on grassland. It also explains the effect of increased atmospheric concentration and climate interaction. This chapter also sheds light on the significance of climate change effects on forests and grasslands.

The fourth chapter introduces the readers to the concept of environment and crop production. This chapter also explains the crop production and environmental protection, heavy metal stress and crop productivity, effect of industrial pollution on crop productivity.

The fifth chapter throws light on the several challenges and progress of wheat production. This chapter contains different challenges of maintaining wheat production, prospects of doubling global wheat yields, creating and exploiting new diversity, and the importance of considering food security in addressing challenges of wheat production.

The sixth chapter takes the readers through the concept of climate change and food security with emphasis on wheat. The readers are then told about the various ways farmers can adapt to climate change and generate income, agriculture's role in addressing a changing climate.

The seventh chapter explains the breeding strategies and quality enhancement under climate change. This chapter also emphasizes crop adaption to overall extreme climate stresses, exploring genetic resources to increase adaptation wheat to climate change, and breeding methods and approaches to increase genetic gains.

In the last chapter of this book sheds light on crop adaptation to climate change. This chapter also mentions about habituating agriculture to climate change, impacts of climate change and adaptation strategies for agronomic crops.

This book has been designed to suit the knowledge and pursuit of the researcher and scholars and to empower them with various aspects of the production of wheat, so that they are updated with the information. I hope that the readers find the book explanatory and insightful and that this book is referred by the scholars across various fields.

Chapter 1

Wheat Production: An Overview

CONTENTS

This chapter deals with the genetics, naming, and the detailed history of wheat, how it started from the era of landrace, to its best selection and inheritance. It also discusses about the various types of wheat that are cultivated worldwide. It tells about the infections, pests, and diseases that damages the wheat crop and also about the crop breeding methods for wheat improvement. It highlights how the traditional strategies are prepared and used for the utilization of wheat.

This chapter foregrounds the ecology and climatic requirements, the role of environment into the sowing of wheat. The favorable conditions needed for its production, along with discussing the cultivation harvesting and threshing of the wheat, the seasons suitable for doing so. It shifts the focus to the uses and importance of wheat in everyone's life. Towards the end it describes the economic importance and taxonomy of wheat including the cultivation and uses of the wheat crop and the wheat genome.

1.1. INTRODUCTION

Developed everywhere on the world, wheat covers a greater amount of the world's surface than some other grain crop. Nonetheless, despite the fact that it takes more land space than different cereals, in view of a long-term normal, it is simply the third-biggest grain crop, behind maize and rice.

The domestication of grains and the improvement of rural ways of life prompted critical changes in individuals' lives, empowering lasting settlements, the advancement of progress, and exchange. Wheat's domestication created bigger grains and a more beneficial yield, which couldn't have made due in the wild and required proceeded with mediation of farmers purposefully planting it.

As one of the principal grains to be domesticated, present day wheat created from development beginning in the center east around 9,000–11,000 years prior in the rich bow of the center east. Without a plainly recognizable time span, the Neolithic time frame is distinguished by the domestication of yields and animals, which started with the improvement of cultivating and suffered until the advancement of metal devices.

By 4,000 BC the extending topographical scope of cultivating brought about bread wheat turning into a typical staple from England to China. Despite the fact that rice was more essential to the advancement of East Asian societies, wheat was the nourishing establishment for societies in Europe, the center East and western Asia.

Figure 1.1. Wheat production.

Source: Image by SeedQuest.

Wheat was presented in Mexico by the Spaniards around 1520, and to early American pioneers in the 1600's. Around then, it was not well known in New England because of the dirt's and environment, however during the 1800's wheat was developed from seeds presented by moving Europeans and horticultural researchers in the space that would later be known as the "Wheat Belt."

The 1830's saw the advancement of the procuring and sifting machines, permitting farmers to incredibly build their profitability during harvest. The advancement of the steam motor in the 1880's and the inner ignition motor in the 1920's expanded rancher efficiency during both planting and collect, and therefore wheat fields increased.

Wheat is by and large become expected for nourishment for people, however lesser quality wheat and the supplement thick results of flour refining are used for creature feed. Wheat is additionally used for bourbon and lager creation, and the husk can be isolated and ground into grain. Prior to the introduction of corn into Europe, wheat was the chief source of starch for measuring paper and cloth.

Most breads, even rye and oat breads, are made with at least a portion of wheat flour on account of two fundamental quality attributes of wheat that improve the breads – its gluten, and its alpha-amylase activity. High gluten flours offer flexibility in the mixture, considering it to ascend without growing huge air pockets. Delicate cakes, similar to pie outsides and bread rolls, are best with low gluten flours.

All wheat flours are best with low alpha-amylase activity, since alpha-amylase goes starch to sugar and forestalls improvement of appropriate mixture qualities. Numerous flours are painstakingly mixed combinations of both hard and delicate wheats planned decisively for a particular reason.

Wheat cultivars can be classified by planting season, hardness of the grain, and (less significantly) shading. Winter wheats are winter tough, so they are planted in the fall. In the spring they continue development and are reaped right off the bat in the late spring. Spring wheats are planted in the spring and collected late in the mid-year. Spring wheat yields are fundamentally lower than winter wheat yields, yet it offers an extremely great quality for bread making. Delicate wheat assortments have bland pieces (less gluten) which factory simpler than the hard wheats. Delicate wheat flour is liked for piecrust, French bread, rolls, and breakfast food sources. Hard wheats have higher protein and gluten levels than their gentler cousins and are used for bread, cakes and flour.

The hardest wheat is durum (*T. durum*), whose flour is used in the assembling of macaroni, spaghetti, and other pasta items. Delicate wheats and white wheats for the most part bring more exorbitant costs since they are simpler to process and don't need blanching.

Wheat, any of a few types of cereal grasses of the class Triticum (family Poaceae) and their eatable grains. Wheat is one of the most established and generally significant of the cereal harvests. Of the great many assortments known, the most significant are normal wheat (*Triticum aestivum*), used to make bread; durum wheat (*T. durum*), used in making pasta (wholesome glues) like spaghetti and macaroni; and club wheat (*T. compactum*), a gentler sort, used for cake, saltines, treats, cakes, and flours. Moreover, some wheat is used by industry for the creation of starch, glue, malt, dextrose, gluten, liquor, and different items.

The wheat plant has long, thin leaves and stems that are empty in many assortments. The inflorescences are made out of changing quantities of moment blossoms, going from 20 to 100. The blossoms are borne in gatherings of two to six in structures known as spikelets, which later serve to house the resulting a few grains delivered by the blossoms.

In spite of the fact that developed under a wide scope of environments and soils, wheat is best adjusted to mild areas with precipitation somewhere in the range of 30 and 90 cm (12 and 36 inches). Winter and spring wheat are the two significant kinds of the yield, with the seriousness of the colder time of year deciding if a colder time of year or spring type is developed.

Winter wheat is constantly planted in the fall; spring wheat is for the most part planted in the spring yet can be planted in the fall where winters are gentle.

The nutritional composition of the wheat grain changes fairly with contrasts in environment and soil. On a normal, the part contains 12% water, 70% carbs, 12% protein, 2% fat, 1.8% minerals, and 2.2% unrefined strands. Thiamin, riboflavin, niacin, and modest quantities of nutrient and are available, yet the processing measures eliminate the vast majority of those supplements with the grain and germ. Most wheat used for food requires handling. The grain is cleaned and afterward adapted by the expansion of water so the part separates appropriately. In processing, the grain is broken and afterward went through a progression of rollers. As the smaller particles are filtered out, the coarser particles pass to different rollers for additional decrease.

Around 72% of the processed grain is recuperated as white flour. Flour produced using the entire piece is called graham flour and gets foul with delayed capacity in view of the germ-oil content held. White flour, which doesn't contain the germ, protects longer. Mediocre and surplus wheat and different processing results are used for domesticated animals to take care of. The best segment of the wheat flour created is used for breadmaking. Wheat filled in dry environments are by and large hard sorts, having protein substance of 11–15% and solid gluten (flexible protein). The hard sort produces flour most appropriate for breadmaking. The wheat of humid areas are softer, with protein substance of around 8–10% and powerless gluten. The gentler kind of wheat produces flour appropriate for cakes, saltines, treats, and baked goods and family flours. Durum wheat semolina (from the endosperm) is used for making pastas, or alimentary pastes.

1.1.1. Genetics

Wheat Genetics is more muddled than that of most other domesticated species. Some wheat species are diploid, with two arrangements of chromosomes, however many are steady polyploids, with four arrangements of chromosomes (tetraploid) or six (hexaploid).

1. Einkorn wheat (*T. monococcum*) is diploid.
2. Most tetraploid wheats (for example, emmer and durum wheat) are gotten from wild emmer, *T. dicoccoides*. Wild emmer is the aftereffect of a hybridization between two diploid wild grasses, *T. urartu* and a wild goatgrass, for example, *Aegilops searsii* or *Ae. speltoides*. The hybridization that shaped wild emmer happened in the wild, some time before taming.

3. Hexaploid wheats advanced in ranchers' fields. Either tamed emmer or durum wheat hybridized with one more wild diploid grass (*Aegilops tauschii*) to make the hexaploid wheats, spelt wheat, and bread wheat (Hancock, 2004).

1.1.2. Naming

There are numerous plant arrangement frameworks used for wheat species. The name of a wheat species from one data source may not be the name of a wheat species types in another. Inside a species types, wheat cultivars are additionally grouped by wheat raisers and farmers as far as developing season, for example, winter wheat as opposed to spring wheat, by gluten content, like hard wheat (high protein content) versus delicate wheat (high starch content), or by grain tone (red, white, or golden) (Bridgwater 1966).

Major developed types of wheat:

1. Common wheat or Bread wheat – (*T. aestivum*). A hexaploid species that is the most broadly developed on the planet.
2. Durum – (*T. durum*). The just tetraploid type of wheat broadly used today, and the second most generally developed wheat.
3. Einkorn – (*T. monococcum*). A diploid species with wild and developed variations. Tamed simultaneously as emmer wheat, however never arrived at a similar significance.
4. Emmer – (*T. dicoccon*). A tetraploid species, developed in antiquated occasions however not, at this point in far and wide use.
5. Spelt – (*T. spelta*). Another hexaploid species developed in restricted amounts.

1.1.3. Hulled Versus Free-Threshing Wheat

The four wild types of wheat, alongside the trained assortments einkorn, emmer, and spelt have bodies. This cruder morphology comprises of hardened glumes that firmly encase the grains, and, in domesticated wheats, a semifragile rachis that breaks effectively on threshing. The outcome is that when threshed, the wheat ear separates into spikelets.

To get the grain, further preparing, like processing or beating, is expected to eliminate the frames or husks. Interestingly, in free-threshing (or bare) structures like durum wheat and normal wheat, the glumes are delicate and the rachis intense. On threshing, the refuse separates, delivering the grains.

Hulled wheats are frequently put away as spikelets in light of the fact that the hardened glumes give great assurance against vermin of put away grain.

1.2. PLANT BREEDING

In conventional rural frameworks, wheat is frequently developed as landraces, casual rancher kept up populaces that regularly keep up significant degrees of morphological variety. Despite the fact that landraces of wheat are not, at this point, filled in Europe and North America, they keep on being significant somewhere else.

The causes of formal wheat rearing lie in the nineteenth century, when single-line assortments were made through the choice of seed from a solitary plant noted to have wanted properties. Present day wheat reproducing created in the main long periods of the twentieth century and was firmly connected to the improvement of Mendelian hereditary qualities.

The standard technique for rearing innate wheat cultivars is by intersection two lines using hand weakening, at that point inbreeding the descendants many (at least 10) ages before discharge determinations are recognized to delivered as an assortment or cultivar.

F1 hybrid wheat cultivars ought not be mistaken for wheat cultivars getting from standard plant rearing. Heterosis or half breed power (as in the natural F1 mixtures of maize) happens in like manner (hexaploid) wheat, yet it is hard to deliver seed of crossbreed cultivars on a business scale as is finished with maize, since wheat blossoms are finished and typically self-fertilize.

Figure 1.2. Plant bleeding.

Source: Image by Pixabay.

Business cross breed wheat seed has been delivered using substance hybridizing specialists, plant development controllers that specifically meddle with dust improvement, or normally happening cytoplasmic male sterility frameworks. Half breed wheat has been a restricted business achievement, in Europe (especially France), the USA and South Africa.

1.2.1. Production

Wheat typically needs somewhere in the range of 110 and 130 days among planting and reap, contingent on environment, seed type, and soil conditions. Harvest the executives' choices require information on the phase of improvement of the yield. Specifically, spring compost applications, herbicides, fungicides, and development controllers are commonly applied at explicit phases of plant advancement.

Information on stages can be useful in distinguishing times of higher danger in a given environment. For instance, during the meiosis stage, wheat is incredibly vulnerable to low temperatures [under 4°C] or high temperatures [more than 25°C].

Ranchers additionally advantage from knowing when the banner leaf (last leaf) shows up, as this leaf represents around 75% of photosynthesis responses during the grain-filling period, and therefore ought to be safeguarded from illness or creepy-crawly assaults to guarantee a decent yield.

A few frameworks exist to distinguish crop stages, with the Feekes and Zadoks scales being the most generally used. Each scale is a standard framework that depicts progressive stages came to by the yield during the agrarian season.

1.2.2. Diseases

Evaluations of the measure of wheat production lost inferable from plant infections change somewhere in the range of ten and 25% in Missouri (Palm 1993). A wide scope of life forms contaminate wheat, of which the most significant are infections and growths. Some regular infections incorporate grain yellow bantam infection, wheat streak mosaic (WSM), and wheat soil-borne mosaic (WSBM).

1.2.3. Pests

Wheat is used as a food plant by the hatchlings of some Lepidoptera species including The Flame, Rustic Shoulder-hitch, Setaceous Hebrew Character, and Turnip Moth.

1.3. BRIEF HISTORY OF WHEAT

The greater part of us eats grains each day – in bread, oats, rolls or pasta. Lately, with gluten narrow mindedness on the ascent, wheat has been getting awful press. In any case, what amount do you think about this grain that structures a particularly critical piece of our eating routine, and how has the wheat we eat changed throughout the long-term?

1.3.1. The Era of the Landrace

Wheat has been developed for over 10,000 years, starting in the Fertile Crescent and showing up in the UK around 5,000 years prior. Processing wheat for flour just got normal in the twelfth century, however by the turn of the nineteenth century, wheat was the UK's most huge yield developed for human utilization.

Nonetheless, this wheat was totally different to the harvests that fill our fields today – the ears would overshadow our cutting-edge bantam assortments, usually arriving at 160 centimeters tall, and with extraordinary hereditary variety. These 'landrace' assortments were made by ages of characteristic choice and ranchers saving different seed a seemingly endless amount of time after year.

Over the long haul, the landrace would get adjusted to the particular soil and environment of the district, as the genotypes that do best in those conditions turned out to be more predominant. Notwithstanding, the quest for better returns and industrialization of farming in the course of recent years has implied these antiquated assortments have been lost from our fields and all that survives from these conventional landraces is a small bunch of seeds that make up a progression of sections, known as promotions, in quality banks all throughout the planet.

Figure 1.3. Brief history of wheat.

Source: Image by Wikimedia Commons.

1.3.2. Selecting the Best

"Around the mid-1800 s, the principal plant breeders understood that on the off chance that they saved the best ears out of the landrace, they got single assortments that yielded higher, however weren't as different," clarifies Ed Dickin, a sharp grain raiser who addresses in crop creation at Harper Adams University.

The first of these 'seedsman's' assortments, for example, Square heads Master, were created during the 1860 s; estimating in at around 130 centimeters, they had a more limited and stiffer straw just as an essentially better return.

At a comparative time, around the 1870 s, the UK started bringing in more wheat from Canada. "The mill operators began using roller plants, and roller processing functions admirably with a hard wheat with a thick grain, as the wheat is effectively isolated from the white flour stream," Ed clarifies. This, joined with the rising interest for white portions to make the sandwiches of the undeniably industrialized labor force, implied bread cooks were additionally excited about the higher protein levels of the imported wheat.

1.3.3. The Laws of Inheritance

Around 1900, crafted by the priest Gregor Mendel was rediscovered. Mendel had dealt with the thoughts of genetics and the laws of inheritance at a comparative chance to Darwin, yet it wasn't until the turn of the twentieth century that this work was applied to the rearing of wheat.

It was found that albeit imported Canadian grains didn't deliver a decent yield in our environment, they actually created high protein grain, and the recently shaped Plant Breeding Institute (PBI) set about catching this quality by intersecting it with a portion of the British assortments.

A vital piece of wheat genetics is that the plant is self-pollinating, which means the dust from the anthers falls onto the disgrace inside a similar bloom. To cross assortments, raisers should eliminate the male anther a piece of a wheat blossom before it produces dust, at that point, once the shame has developed, present dust from the plant they wish to cross the wheat with.

This cycle, known as hybridization, delivers an original F1 plant that will be a hereditary cross of the guardians. Notwithstanding, the future, a F2, will have immense variety as a result of the enormous number of genotypes made by the hybridization interaction. To create a steady assortment, various ages of self-fertilization and cautious choice of plants is required. This is the strategy the PBI used in 1916, crossing the Canadian Red Fife with a low protein British assortment to deliver Yeoman, a hard wheat that deliberate in at around 110 centimeters tall. While this rearing improved the bread making properties of local wheat, the UK was all the while bringing in a lot of its bread wheat – an attribute that proceeded until the 1960 s when a triplet of changes came into place. The previously was the appearance of the Chorleywood Bread Process, an industrialized interaction that could use lower protein wheat. In blend with this, the Plant Varieties and Seeds Act of 1964 permitted breeders to gather sovereignties from the seeds they bred, and the UK's entrance into the European Economic Community (EEC) implied that there were duties on the import of Canadian wheat – therefore, the utilization of British wheat started to increase.

1.3.4. The Quest for Higher Yields

Since the 1980 s, raisers of processing wheat have been progressively cantered around their biggest market – white cut bread – implying that the advanced assortments, for example, Skyfall, are bred for this reason, delivering exceptional returns, of adequate protein levels for roller processing and the Chorleywood Bread Process.

Bread cook, mill operator and grain cultivator Andy Forbes bring up the more extensive effect this had on the wheat developed: "Wheat used to be tall to out contend the weeds, yet in the event that you put synthetic compost on it to expand the yield, there's a danger it will fall over, known as housing. That is one of the principal reasons more limited assortments were developed, however whenever it's abbreviated, the wheat doesn't conceal out the weeds, so farmers began using herbicides to dispose of weeds."

One of the worldwide figures at the core of building up these new wheat was Norman Borlaug, who led the supposed 'green upset.' Yet, while the grain volume created by his semi-bantam harvests was essentially higher, the utilization of nitrogen compost to drive up yield, and herbicides and fungicides to oversee illness, an affects soil wellbeing, and the hereditary homogeneousness of current wheats additionally implies they're exceptionally powerless to sickness.

"By developing the unadulterated line assortments, you get a high return, yet you lose the capacity to adjust," Ed clarifies. "The yellow rust populace, similar to some other illnesses, is a different populace, so the pieces of the populace that can beat the plants' obstruction become more prevailing. Be that as it may, the cutting-edge wheat can't adjust in light of the fact that it's a monoculture, so the raisers are continually getting new wellsprings of protection from address this. It's the raisers' treadmill."

The pressing factor for high yielding wheats is likewise determined by the manner in which the product market works. Most grain in this nation is sold by means of a broker, who goes about as an agent among farmers and purchaser – regardless of whether that is a flour factory, grain exporter or feed plant.

The merchants set the cost dependent on the worldwide market, removing the deal cost from farmer's control. "Delivering a ware implies the expense of creation is the solitary thing a farmer can sensibly hope to impact," farmer Fred Price from Gateley Farm brings up. "As a cost taker, this makes an unavoidable predisposition towards expanding yield trying to diminish cost of creation."

1.3.5. Fit for Purpose?

From the agroecological farmer's viewpoint, current wheat don't fill well in a low information, natural framework; just, they're intended for an alternate developing framework that is dependent on outside inputs. "Natural farmers don't approach or need to use these synthetic substances," Andy clarifies.

"Yet, they additionally don't approach wheat that needn't bother with these sources of info, or that are fit to their neighborhood environment."

Furthermore, there has been an expansion sought after for sourdough bread, just as an ascent in mindfulness about the medical advantages of entire grains. However, the advanced wheat we fill in this nation is fit to neither market.

"Thick brenned wheat that is reasonable for roller wheat is nothing but bad for stone processing," Ed brings up. "You need an assortment with a slenderer wheat since all the grain will go in the flour." Similarly, the gluten levels in the flour will be not the same as those used in Chorleywood bread, which means these flours aren't ideal for sourdough baking.

1.4. TYPES OF WHEAT

Most of individuals believe that all flours are something very similar, quite often produced using wheat or from without gluten sources like rice. Actually, there is a wide assortment of grains that can be ground with a home plant for various utilizations, with altogether different final results.

The fundamental sorts of wheat accessible to American shoppers are recorded underneath; while there are a few substrains and assortments inside every classification that make for unobtrusive contrasts in the food created, the general characteristics of every wheat type will in any case radiate through.

Figure 1.4. Types of wheat.

Source: Image by Wikimedia Commons.

1.4.1. Red Wheats

First on the rundown is Hard Red Winter Wheat, perhaps the most generally developed and used in the United States. It is developed frequently in Kansas and different Plains states, and its moderate attributes make it exceptionally well known.

Since Hard Red Winter Wheat has a protein substance of roughly 10.5%, it is frequently used by business processes in universally handy blends.

It is similarly well known for the individuals who own grain factories and use the flour for hard moves, Asian-style noodles, and container heated breads.

Next in the rundown of basic wheat is Hard Red Spring Wheat. This assortment is developed regularly in the northern states and southern Canada and is one of the hardest wheat assortments used in America.

It is likewise known for its high protein substance of 13.5%, the most noteworthy of all wheat assortments.

Despite the fact that it is one of the more troublesome grains to crush, it is as yet used to make croissants, pizza batter, and delicate breads of a reliably great contrasted and different flours. Therefore, numerous dough punchers add a limited quantity of spring wheat to their flour blends; this builds the protein content and furthermore makes Hard Red Spring wheat a typical mixing specialist.

Delicate Red Winter Wheat comes third in the rundown. This delicate wheat is low in protein and is frequently used for making pasta, dish bread, cakes, or grain. Since it is delicate, it isn't as hard to crush and hence is a decent decision for the individuals who have a more modest granulating cone on a manual grain factory, for example, the Family Grain Mill or the Victorio Deluxe.

At the point when ground at home in a grain factory, each assortment of red wheat has an unmistakable flavor and shading. Regularly considered as having a "good" or "nutty" desire for examination with conventional universally handy (white) flour, Hard Red Wheat of both winter and spring assortments is an awesome expansion to a family's storeroom or crisis food stockpiling supply.

Then again, delicate wheat is by and large not used for crisis stockpiling on account of its low protein content and is best kept in the storeroom for regular use.

1.4.2. White Wheats

Hard White Winter Wheat is the fourth to consider and is a generally new wheat assortment, however it is getting more well known. Its protein content is like Hard Red Wheat, yet as one would figure from its name, it isn't red and has a more unbiased, sweet flavor.

Numerous buyers favor Hard White Winter Wheat since it actually has the very rich flavor that comes from home crushing while at the same time bringing about bread substantially more like locally acquired white bread. This wheat is used for large numbers of similar items as Hard Red Winter Wheat: skillet breads, Asian-style noodles, etc.

The fifth wheat in our rundown is Soft White Spring Wheat, which is filled fundamentally in California and in the east. It shares numerous attributes for all intents and purpose with Soft Red Winter Wheat, and cooks will in general use it for cakes and baked goods that need a fairly better flavor and lighter tone.

In the first place, what is essentially ordered by its developing season.

1. Winter Wheat – planted in the fall, collected in the spring, involves around 75% of wheat-filled in the U.S.
2. Spring Wheat – planted in the spring, collected in prefall or late-summer.

   Past the developing season, wheat is additionally arranged by its hardness (hard/delicate), shading (red/white) and state of its part. Eventually, we're left with the accompanying six kinds of wheat:

1. Hard Red Winter
2. Hard Red Spring
3. Soft Red Winter
4. Durum
5. Hard White Wheat
6. Soft White Wheat

1.4.3. Hard Red Winter

Hard Red Winter has its foundations immovably planted in the Great Plains (Texas, Oklahoma, Kansas, Colorado, and Nebraska) and is dispersed somewhat in different states, too. It's an amazingly adaptable wheat that is extraordinary for bread and generally useful flour.

1.4.4. Hard Red Spring

Hard Red Spring comes basically from the Northern States (Montana, South Dakota, North Dakota, and Minnesota). Now and then alluded to as the "blue-blood of wheat," it's extraordinary for fashioner breads, bagels, croissants, and then some.

1.4.5. Soft Red Winter

Soft Red Winter is generally discovered east of the Mississippi River. For you treat beasts out there, this is the wheat for you! It's extraordinary for treats, saltines, pretzels and cakes!

1.4.6. Soft White

West Coast address! soft White Wheat calls the Pacific Northwest (Washington, Oregon, and Idaho) its home. This wheat makes mouth-watering cakes, cakes, and Middle Eastern flatbreads.

1.4.7. Hard White

This is the most current class of U.S. wheat! Hard white offers its domain with Hard Red Winter wheat, however is developed on a lot more limited size. This class is incredible for Asian Noodles (ramen, anybody?) tortillas and flatbreads.

1.4.8. Durum

Durum is for pasta darlings out there! It's filled essentially in Montana and North Dakota with a smidge filled in Arizona, California, and South Dakota. This wheat has a high protein (also known as gluten) content that is perfecto for pasta and couscous.

1.5. PREPARING: TRADITIONAL STRATEGY

Reaping → Threshing → Winnowing → De-stoning (arranging to eliminate stones and unfamiliar materials → Cleaning in water to eliminate following soils → Draining → De-hulling in mortar → Winnowing to eliminate wheat → Washing de-structure grains → Sun-drying → Milling.

1.5.1. Utilization

100 grams of hard red winter wheat contain about 12.6 grams of protein, 1.5 grams of all-out fat, 71 grams of starch (by contrast), 12.2 grams of dietary fiber, and 3.2 mg of iron (17% of the day-by-day necessity); a similar load of hard red spring wheat contains about 15.4 grams of protein, 1.9 grams of all-out fat, 68 grams of sugar (by distinction), 12.2 grams of dietary fiber, and 3.6 mg of iron (20% of the day-by-day prerequisite).

A large part of the carb part of wheat is starch. Wheat starch is a significant business result of wheat, yet second in financial worth to wheat gluten. The chief pieces of wheat flour are gluten and starch. These can be isolated in a sort of home trial, by blending flour and water to shape a little chunk of batter, and working it tenderly while washing it in a bowl of water. The starch drops out of the mixture and sinks to the lower part of the bowl, leaving behind a chunk of gluten.

Wheat grains have been used by man for quite a long time. Different dishes have advanced over the long run (Olabanji et al., 2004). Around 2/3 of the wheat developed overall is used as food, while the rest of used as feed and as seed for planting (about 7% of developed wheat).

Wheat starches are used mechanically for paper estimating, clothing starch, glucose and glucose syrup creation, assembling of board, glues, glasses, dormant transporters for synthetic compounds, creation of liquor, plastics, disappears, cleansers, elastic and beautifying agents.

1.6. THE DIFFERENT PURPOSES OF DIFFERENT TYPES OF WHEAT

The protein substance of the wheat is the thing that basically figures out what the wheat will be used for and the more noteworthy the protein, the more prominent the versatility of the mixture will be.

1. Hard red winter and hard red spring contain the most noteworthy rates of protein are frequently used in merchandise requiring size, similar to breads and rolls.
2. Hard white and soft white contain the most reduced rates of protein and most appropriate for heated merchandise like cakes, treats, saltines, cakes, and biscuits.

The shade of the wheat has an influence as well, and as you may have effectively speculated, red wheat is hazier than white wheat. Red wheat additionally has a more grounded, harsher flavor than white. This is certainly not a serious deal for a significant number of us home cooks, yet it is serious when you're a major maker attempting to make an item that is outwardly interesting to your shopper.

1. Soft red winter is frequently used in unmitigatedly clear earthy colored wafers and level breads. These are frequently showcased as "entire grain" saltines and such.
2. Soft white is used in merchandise when produces need the thing to look and taste "white," however have the option to guarantee as solid with wheat. One illustration of this are biscuits "made with white wheat."

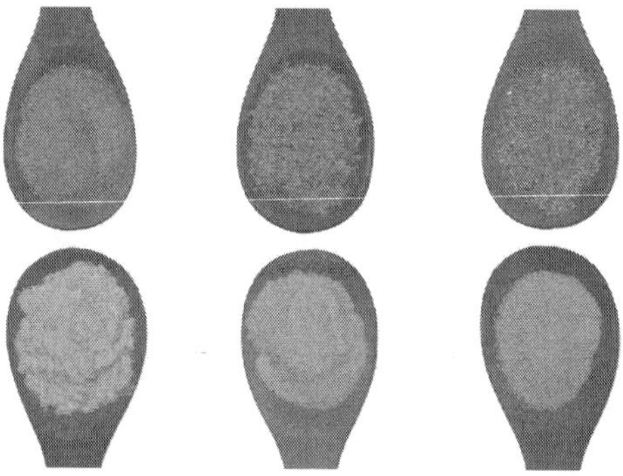

Figure 1.5. The different purposes of different types of wheat.

Source: Image by Pixabay.

1.7. MIXING DIFFERENT TYPES OF WHEAT FOR DIFFERENT FLOURS

As a rule, the healthful profiles of the grains don't vary much past the protein. Basically, you could trade one wheat for another and the sustenance (with the exception of the protein) would remain genuinely steady. That is the reason one will discover numerous individuals using white wheat for bread at home when they're attempting to wean their families from handled white flour.

Be that as it may, the possibly time you're getting 100% entire wheat is the point at which the mark in a real sense says "100% entire wheat" or "100% white entire wheat," in which case the solitary contrast is the shading, reaping season and protein content.

There are numerous different flours that aren't 100% entire wheat, however. So, at that point, what are they made of?

Figure 1.6. Mixing different types of wheat for different flours.

Source: Image by Flickr.

1.7.1. All-Purpose Flour

80% hard red wheat, 20% delicate red wheat. Recollect that this is made distinctly from the endosperm, which has almost no (assuming any) supplements and next to no shading, since the germ and wheat are normally more obscure. Recall that All-Purpose flour will in general be dyed and advanced.

1.7.2. Bread Flour

Most assortments are produced using hard red spring wheat, since it contains the most elevated level of protein and bread flour is ideal for bread. This is likewise processed uniquely from the endosperm, so there is close to nothing, assuming any, nourishment. Bread flour likewise will in general be blanched and improved.

1.7.3. Cake Flour

Typically got from soft white wheat, yet again just from the endosperm.

1.7.4. Pastry Flour

With a somewhat higher level of protein than cake flour, this is likely gotten from the endosperm of hard white wheat.

1.7.5. Self-Rising Flour

A mix of generally useful flour, heating powder and salt. Not suggested, since the generally useful flour is likely dyed and there's no power over the amount of preparing powder and salt in the blend (which makes using it in a formula that calls for the two fixings almost inconceivable). Besides, the preparing powder can get inadequate in muggy environments.

1.7.6. Stone-Ground Whole Wheat Flour

Made as our forefathers would have done it with stones granulating the wheat berries rather than steel plants. Bread specialists guarantee stone-ground flour holds a greater number of supplements than steel processed in light of the fact that the warmth created from the steel plants can hurt a portion of the supplements.

1.7.7. Specialty Varieties

The last wheat assortment is the hardest known to man, and furthermore have the most noteworthy measure of protein. Filled transcendently in North Dakota, durum wheat is used to make the greatest noodles and pastas (e.g., Durum wheat is used only in the incredibly famous pasta created in Italy).

On the other hand, Durum wheat has low measures of gluten; this implies that while it is an extraordinary decision for blending in with different flours, not very many plans will use just Durum flour.

These five sorts of wheat structure the premise of a wide assortment of food sources from essential sandwich bread to the most impeccable baked goods. They are likewise the essential elements for the home grain factory client, who profits by having the freshest potential flours for greatest sustenance and flavor in their prepared merchandise.

Sorting the sorts of wheat gets much more unpredictable than recording it in a class. Cutting edge assortments resemble the books that make up the wheat library. Each rancher needs to choose which book to look at before he leaves. Suppose the rancher realizes that he just needs to peruse experience books. He will try not to go to different sorts, yet he needs to do some examination about which book sounds great to peruse.

Or on the other hand, perhaps a rancher likes both experience and sentiment books. They are both in a similar wing of the library, so he can peruse to-and-fro between segments, in any case, the books he will look at are from these two classifications. Some geographic locales are appropriate to numerous classes of wheat, yet farmers need to know which wheat will work best on their homestead.

Farmers' assortment decisions are restricted by the classes fit to their developing district. However, don't let this moron you, they actually have many assortments to look over. Everest is a famous assortment planted in focal Kansas, however TAM 114 is somewhat more qualified for the Texas beg. They are both hard red winter wheat assortments, however they are more qualified to various geographic conditions.

farmers additionally need to consider which plant qualities are appropriate for their fields. Every assortment has various abilities. Some are great at opposing a particular infection that influences yields. Others are better aggregated to specific soils. Some need less downpour to deliver grain than their adjoining cousins. Every farmer needs to settle on a choice, here and there field by field, about which wheat assortment will work for them.

1.8. ECOLOGY AND CLIMATIC REQUIREMENTS

The creation of wheat in Nigeria is inside scope 10–14^0 N and elevation of 240–360 m above ocean level in the Sudan and Sahel savannah. Ideal air temperature for development and the advancement of grain is 18–25°C. High temperatures during formative stages rush development and improvement, while shortening different formative stages, making it difficult to accomplish full yield possibilities of the harvest.

Creation of wheat was at first restricted to the Chad Basin, and since calm assortments were used, its development was kept to the dry cold time frame among November and March, along these lines requiring the utilization of water system.

Environment, as the normal climate state of a spot for in any event 35 years decides, what the rancher can plant, however major social practices and rate of vermin and infections.

In spite of the fact that wheat is known to have an expansive transformation, it best flourish under mild environment. High temperatures by and large breaking point yield of wheat crop.

The significant yield restricting elements of the environment in the jungles are temperature and mugginess. Nonetheless, under dry conditions, high soil temperature can adversely effect on yield, while dampness will support rate of pests and diseases.

Figure 1.7. Ecology and climatic requirements.

Source: Image by Pxhere.

Temperature prerequisite for wheat assumes huge part in site determination for wheat development, notwithstanding temperature necessity, wheat flourishes best in all around depleted fruitful soil sand to medium surface clay loam and space of low night temperature.

Information on stages is imperative to recognize times of higher danger, regarding the environment. For instance, the meiosis stage is amazingly vulnerable to low temperatures (under 4°C) or high temperatures (more than 25°C).

1.8.1. Sowing Date

Sowing date altogether impact crop development, improvement and yield. Despite the fact that exploration to create downpour took care of wheat assortment is on, wheat is fundamentally a calm harvest, requiring mild conditions which just exist in the Sudan and Sahel savannah during the dry virus period of November/December.

Along these lines, the yield is normally planted between November/December to be gathered in March/April. The ideal time for seed planting is mid-November. Planting prior or later than Mid November will influence yield in wheat crop. Deferral in seed planting has been seen to open the yield to high temperatures, aphid and stem drill assaults.

1.8.2. Seed Treatment

Setting up a sufficient yield stand is an essential for an effective wheat crop, consequently seed treatment is a crucial cycle in accomplishing this. Seed treatment will guarantee high-rate germination, great harvest stand; and taking everything into account, great yield.

Seed treatment with fitting pesticide will improve better seed germination by forestalling or diminishing the occurrence of nuisances. Cover Star 50DS applied at the pace of 10 g of Apron Star for 4–6 kg of seeds is the normal practice.

1.8.3. Land Preparation

Land arrangement necessity for wheat relies upon the trimming history of the land to be used. As a dry season crop, land planning generally starts early November after the gather of the last downpour took care of yields. It is be that as it may, normal to have land arrangement done before the finish of October.

The current harvest under development preceding consolidating the wheat crop into the editing framework, will decide when to start the cycle of land planning. The Site ought to be furrowed and harrowed to a decent slant. Since wheat crops are developed underwater system in Nigeria, there is normally no requirement for ridging as water system is by gravity.

Water development by gravitational force requires accomplishing a degree of slant. Evening out the field to an incline of 0.25–0.30% to a channel situated at the tail of the field will guarantee free water development. Depressed beds are set up in such a manner to consider free water development.

1.8.4. Fertilizer Application

Wheat reacts well to NPK fertilizer application. Utilization of 100–120 kgNha-1 and 40–60 Kg P_2O_5 ha^{-1} and 40–60 kg K_2O ha^{-1} has been suggested. P and K ought to be applied preceding seed planting and consolidated into the dirt. Besides in sandy soils, N can be applied preceding seed planting as a single portion. In any case, for profoundly permeable sandy soils, N ought to be applied in split portions, first at planting, at that point three or a month after seed planting.

Current suggestions regularly show that the second utilization of nitrogen be done when the ear (not apparent at this stage) is around 1 cm in size (Z31 on Zadoks scale). A few frameworks exist to recognize crop stages, with the Feekes and Zadoks scales being the most generally used.

Each scale is a standard framework which depicts progressive stages came to by the harvest during the agrarian season. Harvest the executives' choices require the information on phase of advancement of the yield. Specifically, spring compost applications, herbicides, fungicides, development controllers are regularly applied at explicit phases of plant improvement.

1.9. CULTIVATION, HARVESTING, AND THRESHING OF WHEAT

In northern India, planting of wheat is done in October-November. Wheat might be planted transmission, either by hand or by planting machines. Germination starts quickly and the main leaves show up inside a fortnight. Wheat is appropriately manured and irrigated.

The yield is gathered by cutting the plants with a sickle near the ground, in March–April. Sifting is the following interaction, and this includes the detachment of the grain from the spike. Threshing is by and large done under the feet of bullocks or by sifting machines.

Subsequent to sifting the wheat is winnowed and filtered. In Punjab, as of late, the joins are presented for the reason. The consolidates procure, clean, sift, winnow and filter the grains; wheat should be put away in solidly constructed designs, and it should be very much ventilated.

Figure 1.8. Cultivation, harvesting, and threshing of wheat.

Source: Image by Flickr.

1.9.1. Milling of Wheat

In India, in the majority of the engine-driven flour plants, there stays a fixed lower stone whereupon a mobile upper one spins. The grains are dropped in the openings in the upper stone and continuously worked out between the stones which have granulating surfaces cut in transmitting lines. In this, interaction entire grain is used.

The roller cycle of processing is a high level and wonderful interaction. The initial phase in this interaction contains cleaning and scouring. In the wake of making the grains clean, they are altogether washed and scoured. The subsequent stage is treating.

In this interaction, a little water is added, which hardens the grain and forestalls its separating, so it will go missing safely. In the last, the adapted and tempered wheat is submitted to breaking, pounding, and rolling.

The grain is first broken or squashed step by step through a progression of four to six sets of chilled iron break rolls. The outside of the break rolls is made unpleasant by sharp the long way collapsing. The break flour is isolated out by sifters, while the fundamental part goes to the subsequent break.

This interaction is rehashed until five arrangements of rollers have been used. All grain is eliminated during this cycle, and the cleansed material is passed to smooth rollers for conclusive granulation. The eventual outcome is the flour.

1.9.2. Uses of Wheat

There are three principal sorts of flour—maida and atta which are used for different purposes. The flour is used predominantly for making 'bread' and 'chapatis.' The flour is additionally used for making bread rolls, cakes, baked goods and comparative articles. Wheat pieces are used as breakfast food.

Wheat is additionally used in the production of lager and other cocktails. Wheat straw is used for seating seats, stuffing sleeping cushions, and so forth it makes a decent nourishment for domesticated animals. Wheat straw is likewise used as grain.

This makes a staple food in many pieces of the world. Properties of gluten in the grains are with the end goal that It produces bread-stuffs commonly better than those from some other cereal grains by results of wheat processing, like wheat, germ and mediocre establish important feed for stock promptly eaten; valuable feeds are given to supply protein and minerals in which the straw is inadequate. The straw is used as bedding for steers; it is additionally used for cushioning, as in sleeping pads pinnacle pressing delicate merchandise, for covering and numerous different purposes. It could be used likewise for creation of furfuryl alcohol. Straw-mash is used for the assembling of paper, straw-board, and building-facial hair.

Non-feed mechanical employments of wheat incorporate the assembling of starch, modern liquor, malted wheat, and center fastener flour; just little amounts of wheat are used for starch and gluten fabricate. Gram is viewed as a backup for liquor creation. Second-rate flours are used in the arrangement of glues for decorating and employ wood glues, and in iron foundries as a center folio.

Wheat items incorporate stripped wheat; Bulgur, a parboiled wheat item; World wheat like Bulgur, however of lighter tone; Instant or agglomerated flour; Farina or semolina-. Wheat chips Shredded wheat; Puffed wheat; Grape-nuts, arranged from toasted cuts of malted bread; Gluten, used in extraordinary breads; and raw grain, plentiful in nutrient E.

1.10. ECONOMIC IMPORTANCE AND TAXONOMY OF THE WHEAT

Wheat (*Triticum aestivum* L. em Thell.) is the main significant and key cereal yield for most of world's populaces. It is the main staple food of around 2 billion individuals (36% of the total populace). Around the world, wheat gives almost 55% of the carbs and 20% of the food calories devoured universally.

It surpasses in grounds and creation each and every other grain crop (counting rice, maize, and so forth) and is subsequently, the main oat grain yield of the world, which is developed over a wide scope of climatic conditions and the comprehension of hereditary qualities and genome association using atomic markers is of extraordinary incentive for genetic and plant rearing purposes.

The grass family Poaceae (Gramineae) incorporates significant harvest plants like wheat (*Triticum aestivum* L.), grain (*Hordeum vulgare* L.), oat (*Avena sativa* L.), rye (*Secale cereale* L.), maize (*Zea mays* L.) and rice (*Oryza sativa* L.). Triticeae is one of the clans containing in excess of 15 genera and 300 species including wheat and grain.

Wheat has a place with the clan Triticeae (= Hordeae) in the grass family Poaceae (Gramineae) (Briggle and Reitz, 1963) in which the one to a few blossomed spikelets are sessile and substitute on inverse sides of the rachis shaping a genuine spike. Wheats (Triticum) and ryes (Secale) along with *Aegilops*, *Agropyron*, *Eremopyron* and *Haynalidia* structure the subtribe Triticineae.

Linnaeus in 1753 first arranged wheat. In 1918, Sakamura announced the chromosome number sets (genomes) for each generally perceived sort. This was a defining moment in Triticum grouping. It isolated wheat into three gatherings.

Diploids had 14 ($n = 7$), tetraploids had 28 ($n = 14$) and the hexaploids had 42 ($n = 21$) chromosomes. Bread wheat is *Triticum aestivum*. *T. durum* and *T. compactum* are the other significant species. Every one of the three are results of common hybridization among ancestral not, at this point developed financially.

1.10.1. Cultivation and Use of the Wheat Crop

Wheat is an edible grain, one of the most seasoned and generally significant of the cereal harvests. In spite of the fact that developed under a wide scope of environments and soils, wheat is best adjusted to calm areas with precipitation somewhere in the range of 30 and 90 cm. Winter and spring wheats are the two significant kinds of the harvest, with the seriousness of the colder time of year deciding if a colder time of year or spring type is developed.

Winter wheat is constantly planted in the fall; spring wheat is for the most part planted in the spring however can be planted in the fall where winters are gentle. Consequently, today wheat is developed everywhere on the world, with various assortments planted by the different environments.

In 2002, the world's fundamental wheat creating districts were China, India, United States, Russian Federation, France, Australia, Germany, Ukraine, Canada, Turkey, Pakistan, Argentina, Kazakhstan and United Kingdom. A large portion of the as of now developed wheat assortments have a place with hexaploidy wheat (*Triticum aestivum* L.), which is known as basic bread wheat and esteemed for bread making.

The best bit of the wheat flour delivered is used for bread making. Wheat filled in dry environments is by and large hard sort, having protein substance of 11–15% and solid gluten (versatile protein). The tacky gluten of bread wheat ensnares the carbon dioxide (CO_2) shaped during yeast aging and empowers raised mixture to rise.

The hard kind of wheat produces flour most appropriate for bread making. The wheat of moist zones is gentler, with protein substance of around 8–10% and powerless gluten. The milder kind produces flour appropriate for cakes, saltines, treats, baked goods and family flours.

Durum wheat (Triticum turgidum L.), which is the fundamental tetraploid type, is likewise significant, despite the fact that its huge, hard grains yield low gluten flour that is the principal wellspring of semolina reasonable for pasta, couscous, burghul and other Mediterranean nearby final results.

Evidently, no monetarily significant diploid wheat are being developed as a yield anyplace on the planet. Albeit most wheat is developed for human food and around 10% is held for seed and industry (for creation of starch, glue, malt, dextrose, gluten).

Wheat grain contains every single fundamental supplement; piece contains around 12% water, including carbs (60–80% chiefly as starch), proteins (8–15%) containing satisfactory measures of all fundamental amino acids (aside from lysine, tryptophan, and methionine), fats (1.5–2%), minerals (1.5–2%), nutrients (like B complex, nutrient E) and 2.2% rough strands.

1.10.2. The Wheat Genome

The haploid DNA substance of hexaploidy wheat (*Triticum aestivum* L. em Thell, 2n=42, AABBDD) is roughly 1.7 *x* 1010 bp, around multiple times bigger than the Arabidopsis genome, multiple times that of rice and around multiple times that of maize.

The enormous genome of bread wheat has come about because of polyploidy and broad duplications, with the end goal that more than 80% of the genome comprises of dreary DNA arrangements. The normal wheat chromosome is around 810 MB, multiple times that of the normal rice chromosome.

This gigantic size emerges from the beginnings of the wheat genome, of its triple construction (ABD genomes) and the consideration of broad duplications with the end goal that over 85% of the entire genome is made out of monotonous, profoundly methylated arrangements.

Current evaluations of quality numbers in higher plants shift somewhere in the range of 25,000 and 43,000. As per current examinations, impacts of around 30,000 qualities represented the wheat plant phenotype.

1.11. CONCLUSION

In order to live a healthy life, food is an essential requirement of the human body. So, to provide nutrients and nourishment, people consume different types of food wheat being one of the best of the cereal foods that gives nourishment to humans as compared to any other food source. Wheat is the most vital diet constituent because of its adaptability and conversion and then serving as an interesting and satisfying food.

This chapter revolves around the production of wheat, from its genetics to naming that is the different names of a wheat species, it also discusses about the plant breeding. The diseases and pests and how the plant infections affect the species of wheat.

Then it goes on elaborating the history of wheat, the development of wheat over ten thousand years, how it was inherited and discovered. It also sheds light on the different types of wheat that are red wheat, white wheat, soft white, hard white, different purposes for different types of wheat. Then shows the conventional strategies and utilization.

This chapter also depicts the ecological and climatic requirements of wheat, the sowing date, seed treatment, land preparation, and fertilizer application. Then emphasizes upon the cultivation, harvesting, and the threshing of wheat, the milling and uses of wheat and also throws light on the economic importance, and the taxonomy of wheat.

REFERENCES

1. (n.d). *1. Introduction.* [e-Book] Available at: https://sundoc.bibliothek.uni-halle.de/diss-online/04/04H184/t2.pdf (accessed on 7 July 2021).
2. Charles, I O., (2016). *The Wheat Crop.* [online] Available at: https://www.researchgate.net/publication/310458715_THE_WHEAT_CROP (accessed on 7 July 2021).
3. Encyclopedia Britannica, (2021). *Wheat | Production, Types, Nutrition, Uses, & Facts.* [online] Encyclopedia Britannica. Available at: https://www.britannica.com/plant/wheat (accessed on 7 July 2021).
4. Gramene, (n.d). *Wheat Introduction.* [online] Archive.gramene.org. Available at: https://archive.gramene.org/species/triticum/wheat_intro.html (accessed on 7 July 2021).
5. Jordan, H., (2021). *Six Classes of Wheat | Wheat Classes & Varieties |* EatWheat. [online] Eat Wheat. Available at: https://eatwheat.org/learn/kinds-of-wheat/ (accessed on 7 July 2021).
6. New World Encyclopedia, (2021). *Wheat.* [online] newworldencyclopedia.org. Available at: https://www.newworldencyclopedia.org/entry/Wheat (accessed on 7 July 2021).
7. Nikita, D., (n.d). *Wheat: History, Cultivation and Uses.* [online] Biology discussion. Available at: https://www.biologydiscussion.com/economic-botany/wheat/wheat-history-cultivation-and-uses/52873 (accessed on 7 July 2021).
8. Precision Agriculture, (2018). *Six Types of Wheat.* [online] Precisionagricultu.re. Available at: https://precisionagricultu.re/six-types-of-wheat/ (accessed on 7 July 2021).
9. Steph, W., (2019). *A Brief History of Wheat.* [online] Sustainable food trust. Available at: https://sustainablefoodtrust.org/articles/a-brief-history-of-wheat/ (accessed on 7 July 2021).
10. Tiffany, (2013). *The Different Types of Wheat and What They're Used for.* [online] Don't Waste the Crumbs. Available at: https://dontwastethecrumbs.com/the-different-types-of-wheat-and-what-theyre-used-for/ (accessed on 7 July 2021).

Chapter 2

Global Wheat Production

CONTENTS

This chapter briefly discusses about the wheat-growing environment and the day-to-day changes that are occurring, including the effects of chemicals, its root system, and its psychological functions as a whole. Furthermore, it discusses in detail the different variants which are affecting wheat production. This includes root signaling, climatic changes and nitrogen requirement and other physical factors such as temperature, water supply, and its affects with different chemicals such as nitrogen and antioxygen enzymes.

2.1. EFFECT OF PHOSPHORUS ON ROOT SIGNALING OF WHEAT UNDER DIFFERENT WATER REGIMES

On the different stimuli, root signaling is the response of the plant roots like soil structure, soil nutrients, different chemicals, and stress conditions. For the different types of activities, root apical meristems are the major sites in response to changes related to the roots.

For the water and mineral nutrients, the root growth defines the extent to which plant explores soil. In the mechanical impedance root system of individual crop plants may encounter large variations to the root penetration. With different foraging strategies root architecture is a highly plastic and environmentally responsive trait that enables plants to counteract nutrient scarcities.

There are some root specific traits such as to improve resource capture and plant development under water-limited conditions root system architecture, sensing of edaphic stress and root-to-shoot communication can be exploited.

Figure 2.1. Effect of phosphorus on root signaling of wheat under different water regimes.

Source: Image by Pxhere.

On the supply of available nutrients in the rooting media and the root system, the uptake of the nutrients is depend upon. For their adaptation to the environment, the ability of plants to respond appropriately to nutrient availability is of fundamental importance. Nutrients such as nitrate, phosphate, sulfate, and iron act as signals that can be perceived.

Within the root, these signals trigger molecular mechanisms that modify cell division and cell differentiation processes, and on the root system architecture will have a profound impact. Root-hair formation, primary root growth, and lateral root formation are important developmental processes and are particularly sensitive to changes in the internal and external concentration of nutrients. In response to mineral nutrition, there is no doubt that differences occur among spices and cultivators, that is genotypes belonging to the same species.

In crop production, phosphorus plays a vital role, and it is also involved in energy transfer in plants. Without phosphorus, carbon dioxide fixation by plants is not possible. There are many plant physiological functions which are dependent on phosphorus such as the utilization of sugars, starch, photosynthesis, energy storage and transfer.

For the cell division and development of meristematic tissues, it is also a constituent of the cell nucleus, and it is also essential. To increase the strength of cereal straw, resist abiotic stresses, stimulate root development, promote flowering, fruit production, and formation of seed and hasten maturity of the crops phosphorus has been reported.

By mixing it with farmyard manure phosphorus utilization efficiency can be improved to increase the yield of wheat. With the single superphosphate in 1:2 ratio farmyard manure mixed, increases the phosphorus efficiency significantly. To adapt to mineral stresses, it would be advantageous if anyone would select, screen, or improve plants for higher capacity.

In the developing countries, this approach is beneficial like Pakistan where capital input resources are limited. Nutrient efficient crop cultivators are required by farmers in these countries, which perform better or do something better than other cultivars when given a considerable amount of mineral nutrient.

Cereals are facing acute problem of drought and temperature stress. The major environmental factor is the low water availability which limits the crop productivity. Root is the place where plants first encounter drought stress; it is likely that roots may be able to sense and respond to stress conditions.

The most common adverse environmental condition is the drought stress that can seriously reduce crop productivity. For drought-resistant crop plants, the mechanism of drought tolerance and breeding has been a major goal of plant biologists and crop breeders. Under the drought stress, the significant progress has been made in understanding root growth.

However, in the root development, there has been no genetically defined drought-adaptive response. But inhibition of lateral root development is a typical adaptive response of roots to drought stress. To drought stress, physiological and molecular biological studies have documented several plant responses, despite the lack of understanding of drought tolerance mechanisms.

The most important factor affecting world agriculture is the lack of sufficient water. Therefore, in order to improve yield while minimizing damage to the environment, increasing the efficiency of water and nutrient use is essential. to obtain water and nutrients from the soil-plant depends upon the capacity of roots.

To the soil water status, the root respiration, carbohydrates allocation (root: shoot ratio) and grain yield are closely related. During soil drying and improve grain yield, reductions in root respiration and root biomass under severe soil drying can improve drought-tolerant wheat growth and physiological activity, and therefore, in the arid regions, it should be advantageous over a drought-sensitive cultivar.

Therefore, objectives of the study were

1. to examine the effect of phosphorus on root signaling of wheat; and
2. to determine the effect of water stress on root signaling.

   The hypothesis, therefore made, was that there is a significant relationship present between wheat roots and P and also between root and drought stress.

2.2. ROOT SIGNALING

2.2.1. Phosphorus and Root Signaling

After the nitrogen among all essential plant nutrients, phosphorus (P) is the second most abundantly required nutrient element, and it is an important constituent of many structural components of the plants. With respect to its production capacity, it determines the soil quality in the agriculture ecosystem. To increase P use efficiency in soil for successful and sustainable crop production, being scarce and nonrenewable natural resource, which is under the threat of rapid depletion as a result of intensive mining across the world more emphasis is being given.

Over the 2 years, a field experiment was conducted to study the ameliorating effects of P on wheat yield, root cation exchange capacity (CEC) and on different doses of Phosphorus was applied as single superphosphate (SSP). To bloom stage, the application of P increased the root CEC of wheat up to bloom whereas nutrient concentration, uptake, and grain and straw yield were found to increase up to maturity.

Figure 2.2. Phosphorus and root signaling.

Source: Image by Flickr.

The capacity of plant roots to increase their carboxylate exudation at low plant P status is an adaptation to acquire sufficient P at low soil P availability. With the increasing P supply for *Triticum aestivum* L root mass ratio has been decreased. Under sterile conditions, an experiment was set up to make a critical assessment of the role of organic P in soil solution in the nutrition of wheat plants.

On the dry matter, phosphorus supply had a positive effect and the P concentration of the plants. In the organic P treatments, acid phosphatase secretion by plant roots was 5–11 times higher than in the inorganic P treatments. In response to the presence of organic P in soil solution, it was hypothesized that plants secrete phosphatases, and for the increase in P influx to wheat plants, organic P might be responsible.

For the uptake of a less mobile soil nutrient such as P by crop plants root-soil contact is an important factor. Root soil contact can be substantially increased by the root hairs. For increasing P uptake in low input agriculture, the identification of crop cultivars with more and longer root hairs can be useful.

To the quantity of P depleted from rhizosphere, the variation in root hair parameters of the cultivars was related. In root hairs of cereal cultivars, these results showed the variation and can be considerable, and it can play a significant role in P acquisition, especially in low-P soils.

To the low-P stress at the stage of seeding a field trial was conducted to investigate the main morphological and physiological changes of different wheat landraces. Root volume, total leaf area, and plant dry weight is significantly decreased by the P-deficiency but greatly increased density of root hairs and root top ratio.

Moreover, significant enhancement of phosphorus utilization efficiency is being induced by the P-deficiency and the amount of proline, malondialdehyde, acid phosphatase, peroxidase, and superoxide dismutase (SOD), but the significant reduction of P uptake and soluble protein content.

Under the P-deficiency the result based on the correlation analysis showed that the economic yield of wheat landraces had relationships with their morphological and physiological characteristics.

2.2.2. Drought and Root Signaling

The effect of drought on the growth and yield of wheat were investigated. Further yields have been decreased by the drought during grain filling. Due to drought plots with a lower plant density demonstrated a smaller decrease in yield.

Between the number of shoots per plant and nodal root axes per plant, there was a significant positive linear relationship. In the root system architecture, there appeared to be a difference between cultivars and in their response to drought, but these differences were not reflected in grain yield. In the crop yield drought-induced loss which probably exceeds losses from all other causes, since both the severity and duration of the stress are critical.

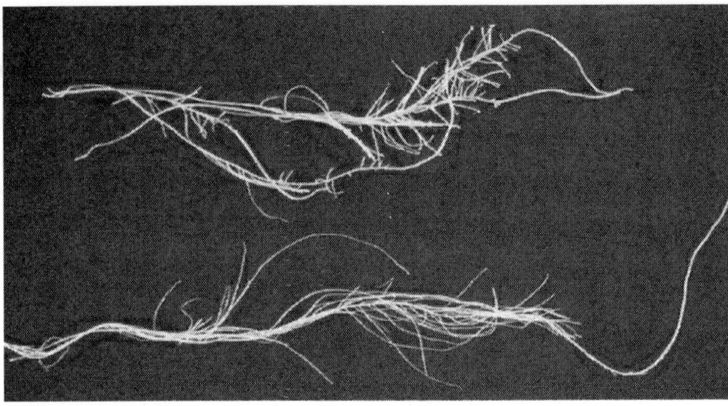

Figure 2.3. Calcium deficiency in wheat.

Source: Image by Flickr.

Drought stress reduces leaf size, stem extension, and root proliferation, disturb plant water relations and reduces water-use efficiency. At cellular plants display a variety of physiological and biochemical responses and whole-organism levels towards prevailing drought stress, thus making it a complex phenomenon. To withstand drought stress plants, display a range of mechanism.

By increased diffusive resistance, the major mechanisms include curtailed water loss, and enhanced the water uptake with prolific and deep root systems and its efficient use, and smaller and succulent leaves to reduce the transpiration loss.

Several drought-responsive genes and transcription factors have been identified at the molecular levels such as the dehydration-responsive element-binding gene, aquaporin, late embryogenesis abundant proteins and dehydrins. There are some of the plant growth substances such as salicylic acid, auxins, gibberellins, cytokinin and abscisic acid modulate the plant responses towards drought.

As an antioxidant's polyamines, citrulline, and several enzymes act and also reduce the adverse effects of water deficit. To increase the yields at higher seeding rates, there is the possibility of reducing the proliferation of roots and at different growing stages conserving the soil water in water-limited environments was studied.

Grain yield decreased as the seeding rate increased in the severe drought towards the end of the growing season, but under the more favorable conditions, the reverse was true. Green yield was significantly increased an average over the seeding rates, grain yield and yield components were higher, and root pruning at spring-growth stage recorded the highest water use efficiency.

Under the drought the leaf net photosynthetic rate and stomatal conductance were significantly decreased. By the drought the leaf transpiration rate was decreased. Under the drought the intercellular CO_2 concentration was increased while it was decreased most of the time from midday to the afternoon. The leaf stomatal limitation was increased under drought.

To select for studying tolerant genotypes under water stress conditions, root length, root dry weight and seedling dry weight are the major traits. During ear formation and flowering stage, it is reported that drought affect the plant water status. Water availability mostly affects the growth of leaves, roots, photosynthesis and dry matter accumulation.

2.3. WHEAT SENSITIVITY TO NITROGEN SUPPLY UNDER DIFFERENT CLIMATIC CONDITIONS

At a faster rate, the global wheat consumption has escalated than all other cereals. By the increase in developing countries, this growth is accounted for, and mainly in China and India, and based on the future projection, the growth of wheat consumption will continue. In these two countries, the use of production inputs, primarily nitrogen fertilizer and irrigation water, has risen dramatically as well.

By people worldwide, wheat is an important staple crop, providing 20% of all calories consumed. In the human food it is the leading source of nonanimal protein and to the animal feed it also makes a significant contribution.

For wheat, increasing global demand is also based on the ability to make several food products and the increasing consumption of these with industrialization. In particular, the properties of the gluten protein fraction allow the processing of wheat to produce bread, other baked goods, noodles and pasta, and a range of functional ingredients.

On plant productivity, the food demand sustainable nutrient supply and climatic effect are the two crucial topics of agricultural development. Based on the genotype requirements applying an adequate amount of nutrients is hard under potential conditions, especially under different abiotic loads.

The amount of yield is being determined by the Nitrogen (N), which is an important nutrient and throughout the proteins the quality as well. With a 20-fold increase in the global use of nitrogen fertilizer during the 50 years the increased crop productivity has been associated, and by the year 2050, this is expected to increase by threefold.

Environmental and ecological problems can be caused by the deficiency and excess of inadequate application of N. By using efficiency and yield, climatic factors can improve and deteriorate the crop nutrient. In the agriculture, drought-induced crop yield reduction is among the greatest losses and drought also occurs in all climatic regions.

In the developing countries, about 32% of wheat production areas experience serious drought stress in different growth stages. In some country's climate trends were large enough to offset a significant portion of the increases in average yields that arose from technology, fertilization, and other improving factors as it is published by Lobell.

On the N use of wheat high and low temperature, irrigation, salinization, agrotechnology, and other nutrients also have an effect. On the adaptation and acclimatization strategies of different wheat genotypes, these effects are depending, and the current climatic conditions and its combinations and biotic effects as well.

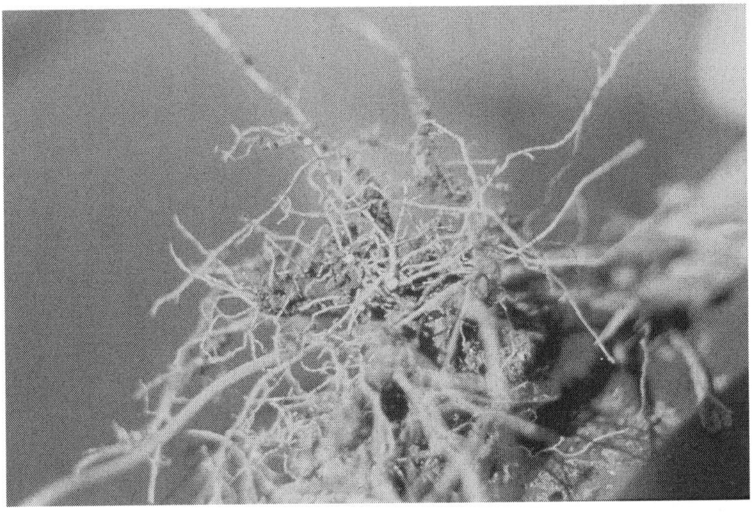

Figure 2.4. Wheat root damage due to cyst nematode.

Source: Image by Flickr.

A way towards the sustainability is being achieved by knowing more about and improve nitrogen use efficiency of wheat. For the qualitative perfect food wheat being the basic food plant and the global demand is increasing, as there are no alternatives, then step forward to smart-wheat, which will be able to survive unfavorable conditions.

2.3.1. Nitrogen Requirement and NUE

In the high quantity, nitrogen is one of the nutrients plants which is needed, as it is a core constituent of a plant's nucleic acid, proteins, enzymes, and cell wall and pigment system. For the plants, the availability of nitrogen is complex and depending on the different processes in connection with nitrogen cycle in the environment. The agriculture has a main impact on global and local nitrogen cycle through the different way of nutrition supply and agrotechnology processes.

By connecting different bacteria or releasing different extracts, like nitrification inhibitors, plants can have an effect on the own nitrogen supply. For the certain plant species, Biological nitrification inhibition (BNI) is the natural ability to release nitrification inhibitors from their roots that suppress nitrifier activity, thus reducing soil nitrification and N_2O emission. It has been said that the BNI function is the strongest in Brach aria, among the tropical pasture grasses.

In the plant life, nitrogen availability and using capacity are crucial. As the photosynthetic machinery accounts for more than half of the N in a leaf, the chlorophyll content of wheat leaves and leaf N is closely related. By leaf growth and leaf area duration and also the photosynthetic rate per unit leaf area the Nitrogen influences carbohydrate source size and thereby source activity.

Because of the reason plant metabolism is differently affected by excess, optimal and deficient levels, the availability of N is of agricultural concern. To characterize plant responses to different levels of N availability, the concept of nitrogen-use efficiency (NUE) has been widely used. Among the breeders, the moll defined the most use of NUE, which computes the grain dry mass divided by the total N available to a plant.

It is divided into two components: NUE = NUpE × NUtE, where NUpE is the N-uptake efficiency calculated as the total amount of N in above-ground plant at harvest divided by the available N in soil, and NUtE is the utilization efficiency calculated as the grain dry mass divided by the total amount of N in above-ground plant at harvest.

As, it is based on the several authors establishment that N remobilization efficiency (NRE) is also a main component of NUE. The ability of plants to translocate the N after anthesis from the shoot to the grains is defined by the proportion of N in the crop or crop component at anthesis, which is not present in the crop or crop component at harvest. For the production of wheat, nitrogen is the most limiting nutrient. Cultivars who are with higher NRE tend to accelerate the senescence process and increase N levels in grains. The major source of the grain N is being provided by the N accumulated before anthesis and it is also widely understood. From the remobilization of N stored in shoots and roots before anthesis, around 50–95% of the grain N at harvest comes in wheat. The leaves had a higher NRE than the stem and the roots, in wheat between anthesis and maturity.

2.3.2. Temperature Influence on Nitrogen Nutrition

By the heat stress wheat growth can be impaired at any developmental stage, and modeling scenarios predict even warmer temperatures in the future. By the high temperature, the production of wheat is being affected. Intensifying N deficiency in plants, as elevated temperature alters uptake and allocation of N. In the canopy architecture, wheat shows the enormous diversity, and it has been proposed that with the use of efficiency as well as light interception, optimized light distribution could improve radiation. The activity of enzymes has crucial role in the heat tolerance. Rubisco's affinity for CO_2 decreases with temperatures. Therefore, increasing affinity would simultaneously improve adaptation to warmer conditions, the proof of concept coming from C4 species, in which it is achieved by concentrating CO_2. By the addition of inhibitory sugars to its active site, high temperature not only degrades Rubisco but also accelerates its inactivation.

Figure 2.5. Temperature influence on nitrogen nutrition.

Source: Image by Flickr.

In addition to it, as compared with the other Calvin cycle enzymes, Rubisco has a relatively low turnover number. By a catalytic chaperone, the activity of Rubisco is mainly regulated and Rubisco activate is the one which catalyzes removal of inhibitory sugars from its active site, switching the enzyme to active mode. Rubisco has one of the best CO_2 affinities among the cereal's wheats. There is a model where wheat's substrate specificity factor of Rubisco is replaced from *L. Gibertii* predicted increases of 12% in net assimilation.

2.3.3. Effect of Drought on Wheat Nutrition

In the year 2014, a report is being published which is based on the most recent assessment of the Inter-governmental Panel on Climate Change, and in history, levels of anthropogenic emissions of greenhouse gases are now at their highest. On the land use, agricultural production and its effect are the major sources of these emissions by sharing methane and nitrous oxide gases.

Air temperature is being increased by the greenhouse gases thus more moisture evaporates from land and water bodies. Warmer temperatures also increase evaporation and evapotranspiration in plants, soils, and on the other hand, they will also escalate the water stress frequency and intensity with a rise from 1 to 30% in acute drought land area by 2100.

In the field 75–100% of the grain yield could be attributed to stored assimilates under the dry conditions, as compared with the 37–39% under high-rainfall conditions. By reducing the water potential drought stress severely influenced plant water status and relative water content in the wheat.

By sustaining metabolic activities under reduced tissue water potential optimal nutrition levels have also alleviated drought stress damage. In combating the drought nitrogen supply also has a crucial role. With the increase in the drought stress, the efficiency of the nitrogen supply has been declined. In terms of growth and development under drought stress plant shows positive influence with an application of nitrogen.

Although, it has been mentioned that under the drought stress different grass species did not modify physiological functions under varying N application. Diffusive conductivity is being reduced by the water limitation, which in turn affects another physiological process such as energy and N metabolism.

It is being concluded that on the environmental condition, N uptake and its diffusion depends, especially to water supply. From the soil, the roots are unable to get the optimal amount of nitrogen under the water deficiency, which has general negative effects on the plant metabolisms.

Due to the marked sensitivity of leaf area expansion, the main effect of water restriction is certainly a reduction in N demand. Mainly under the stress condition, fewer results have about light reaction affected by genotypic and nitrogen supply variations.

2.3.4. Salinization and Impact to Wheat Production

In the soil solution or water resources, the salinization or increased concentration of dissolved cations/anions across the ecosystems is the principal cause of most widespread abiotic constraint to glycophytic known as the salt stress.

As a consequence of primary salinity effects, the salt stress encompasses a wide range of physiological dysfunctions, that is the osmotic and ionic disorder. on the salinity level/duration, crop/genotype type, development stage, primary salinity effects are depending and so on very often cause different secondary salinity-induced effects such as reduced cell expansion and assimilate production (i.e., growth and yield reduction), production of reactive oxygen metabolites and even plant mortality.

There some basic salinity effects that are quite visible and assume reduce biomass growth (shoot/root height and weight, leaf area) and changes in root and shoots color (e.g., presence of leaf tip burns, scorching/firing of leaves). Under salt stress the extent to which growth and yield will be reduced and mostly depends on the salinity level and plant (crop) species.

2.4. ROLE OF OSMOLYTES AND ANTIOXIDANT ENZYMES FOR DROUGHT TOLERANCE IN WHEAT

By recurrent drought situation the global wheat production in the key production areas is being threatened, which is being predicted to increase with the climate change. Against the adverse effects of drought, the drought-tolerant wheat varieties are the ultimate solution of safeguarding the crop.

To the environmental stress, plants are frequently exposed both due to some natural cause and rough agriculture practices. In the limited plant productivity, there are various types of both biotic and abiotic stresses which may result in this.

The result will be in significant crop losses, which are a serious threat to the agriculture, if plant stresses like oxidative, chemical toxicity, drought and salinity, extreme temperatures along with the attack of insects, pests, and plant pathogens.

Crop productivity is being limited when drought is significantly damaging the plant and there many countries that are facing this big disaster. Due to the stressful conditions, plant productivity is greatly inclined that affect almost every aspect of the plant growth.

A unique pattern of biochemical and molecular mechanisms is being developed by all plants to manage the odd situations which are linked with the stress tolerance. The gene expression is being initiated by the environmental stresses like drought, high salinity, and low temperature that raise osmolytes and antioxidant enzyme levels in plant cells to tolerate stress responses.

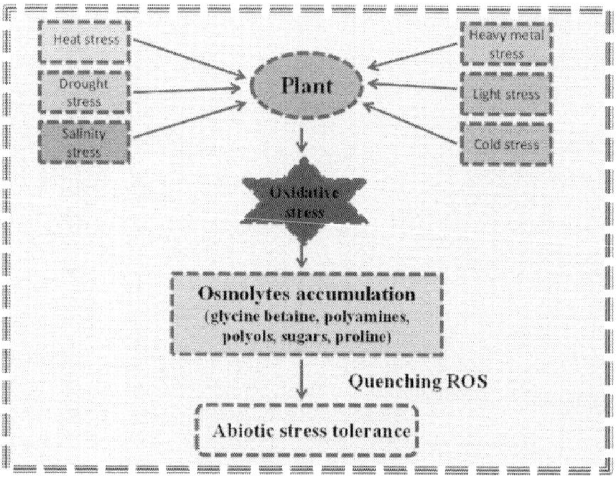

Figure 2.6. Role of osmolytes and antioxidant enzymes for drought tolerance in wheat.

Source: Image by MDPI.

Around the globe, food crop productivity is highly enviable due to ever-increasing population and to expand measures for maximum produce it is the need of the hour. As the staple food wheat occupies an important place and under the different stress there is the yield improvement of the wheat germplasm and for the researchers agro-climatic situations have an essential task to deal with present scenario.

For human consumption worldwide human food comprises many valuable produces like rice, pulses, and meats, but wheat is among the most important. As a feed for the livestock, a part of the total wheat crop production is also being used.

Triticum aestivum is common bread wheat, but the other two species of wheat are of commercial importance as *T. durum* is pasta product wheat and *T. compactum* is pastry flour wheat. As food about 35% of the human population consumes wheat, which is covering 29% of caloric intake. Due to its global production at more than 651. 4 million metric tons per annum wheat shares the largest cereal market.

On the ingredients, the high nutritive value (>10% protein, 2.4% lipids, and 79% carbohydrates) of wheat is based and the versatility of its use in the production of a wide range of food products. Adverse effects are being faced by the global climate change, while some other areas that have adopted the effects of climate change have shown benefits to crop system.

In the region wheat is grown where rainfall ranges 30–113 cm. By the environmental stresses the plant productivity is being hampered, and the plant growth and productivity is limited by the water shortage and even more than any other environmental factor. With the water stress tolerance in crop plants elevated levels of stress hormone viz. ABA has been associated.

Under the desiccation, heat regulates stomatal conductance and water loss that results in the accumulation of osmolytes like proline, mannitol, glycine-betaine, and soluble sugars like trehalose which lower the osmotic potential of the cell sap and thus prevent the movement of water out of the cell.

Due to drought situation Pakistan has faced recently a big problem of wheat shortage. By the drought, the globe is being affected, about 45% of the land area mostly of Africa (Ethiopia) and its surroundings, most of the Mediterranean, Mexico, Australia, and some parts of Middle East, India and Sindh province area of Pakistan.

Irrigated land is only 15% of total cultivated land which yields twice as much as rain-fed land and producing one-third of the world's food. In determining the distribution of plant species, plant stress plays a major role and due to soil texture and climate limitations.

Into the plant to maintain a gradient of water flow, the soil water potential is much important, but its reduction may lead to increased soil solutes that make it increasingly difficult to establish osmotic pressure. In some plant species, osmotic stress leads to stomatal closure and a reduced rate of photosynthesis.

2.4.1. Osmolytes

Based on the synthesis of osmolytes, there are effective mechanisms in plants which are low molecular weight compatible solutes. To accommodate osmotic pressure osmolytes are frequently used by the cells within the effected cells to avoid cellular injury due to oxidation phenomenon. In many organisms, they are highly soluble organic molecules that are synthesized in response to different environmental conditions leading to osmotic stress.

Without interfering, they accumulate in the cytosol with the cellular metabolism even at high concentrations. During the stress response, osmolytes have additional functions and act as osmoprotectants by directly stabilizing protein and membrane structures under dehydration conditions. They have a diverse chemical nature, and apart from contributing to maintain osmotic balance, they do protect cell against oxidative stress as scavengers of "reactive oxygen species" (ROS).

The osmoregulatory such as protein, sugars, amino acids, and compounds of quaternary ammonium play a vital role in adjusting the osmotic pressure and stabilizing of plant cells and tissues. In plants, drought stress causes osmotic stress, which causes a reduction in growth, imbalance ion transport, and a decrease in transpiration rate, and an increase in membrane permeability.

To adverse environmental conditions, such effects result in less water-absorbing capacity of crop plants, and different plant species and genotypes within a species respond differently. Plants accumulate different types of organic and inorganic solutes in cytosol to decrease osmotic potential by which they can maintain cell turgor, in order to counteract unfavorable environmental conditions.

Underwater stress conditions, plant cells lose water and decrease turgor pressure. In the case of water stress like abscisic acid, which has important roles in the tolerance of plants to drought, high salinity, and cold, there is an increase in different plant hormones. For the greatest agricultural losses' abiotic stresses, which cause depletion of cellular water, are responsible.

The accumulation of osmoprotectants is in sufficient quantity to facilitate osmotic adjustment upon exposure to these prevalent stresses. By the influx of water into the cells, there is an increase in the cellular osmolarity due to these compatible solutes, which provide the turgor necessary for the cell expansion. To allow changes in developmental processes, water deficit develops slowly enough, as water stress has several adverse effects on plant growth.

To counter the unfavorable cellular conditions compatible solutes like proline, glycine-betaine, and trehalose produce in this situation. Due to growth arrest and molecular damage the osmotic potential fluctuation of soil solution creating a water stress in plants ultimately leads to plant death. To maintain plant water balance to establish internal milieu, help is being provided by the osmotic adjustments of cell.

2.4.2. Antioxidant Enzymes

For scavenging of reactive oxygen species (ROS) in plants, the enzymatic and nonenzymatic mechanisms are available. To scavenge these harmful ROS by-products of hyperosmotic and ionic stresses causes membrane dysfunction and cell death ultimately the biochemical adaptive function of osmoprotectants provide the support.

These are some active oxygen species that are superoxide (O^{2-}), hydrogen peroxide (H_2O_2), hydroxyl radical (√OH), and singlet oxygen ($1O_2$) which is produced through the oxidation phenomenon. To eliminate superoxide with the help of superoxide dismutase (SOD), many plants have the ability, which catalyzes the superoxide into H_2O_2 and O_2.

In order to eliminate hydrogen peroxide and to save cell membrane from severe damage, thylakoid membrane has potential enzyme ascorbate peroxidase (APX).

2.4.2.1. Superoxide Dismutase

With the degree of stress condition SOD concentrations typically increase, as the compartmentalization of different forms of SOD throughout the plant makes them counteract stress very effectively. In plants, there are three classes of SOD metallic coenzymes that exist and that act to control increased levels of oxidative stress. As a free radical scavenger SOD's role is being established, and those genotypes have higher levels indicating a higher level of stress tolerance in wheat. In the plants, the availability of different forms of SOD shows a maximum stress tolerance in affected crops, giving protection to the plant. So, the trends observed in the present research might be due to the reasons discussed earlier.

2.4.2.2. Catalase

In nearly all living organisms' catalase is a common enzyme which is found and exposed to oxygen and catalyzes the decomposition of H_2O_2 to water and oxygen. By the catalase molecule, the highest turnover could convert millions of molecules of H_2O_2 to water, and oxygen in each second. By-product of many regular metabolic processes, hydrogen peroxide is a toxic. to prevent most cellular damage and tissue injuries, it must be quickly converted into less toxic substances.

2.4.2.3. Ascorbate Peroxidase

By using ascorbate as a substrate ascorbate peroxidase (APXs) are the enzymes that detoxify hydrogen peroxide. To rapidly catalyze the decomposition of hydrogen peroxide into less reactive gaseous oxygen and water molecules, it is frequently used by cells.

During metabolic processes including mitochondrial, chloroplast, and plasma membrane-linked electron transport system, higher plants produce active oxygen species. There some conditions which can give rise to excess concentration of these active oxygen species resulting in oxidative damage at a cellular level due to biotic and abiotic stress.

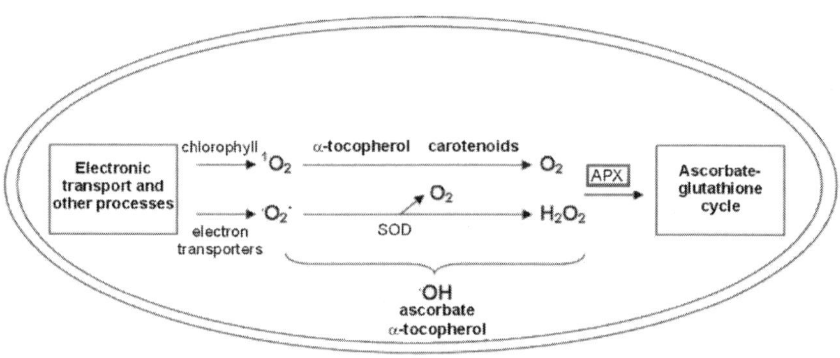

Figure 2.7. Role of ascorbate peroxidase.

To interrupt the cascades of uncontrolled oxidation in each organelle, the antioxidant enzymes have to function at this stage. In the metabolism of H2O2 in higher plants, Ascorbate peroxidase (APX) exists as isoenzymes that play an important role.

Along with activities of other antioxidant enzymes like catalase, SOD, APX activities generally increase and in response to various environmental stress factors regulating the components of ROS-scavenging systems GSH reductase. In many higher plants, APZ has been identified and comprises a family of isoenzymes with different characteristics.

2.5. NITROGEN LOSSES: GASEOUS AND LEACHED NITROGEN BALANCE

In the growth and development of wheat the Nitrogen (N) plays a key role, thus wheat's growth and quality might be modified through N fertilization. However, cultures use N inefficiently, and, in general, 50% of the applied N is not used by plants.

Therefore, N losses take place, both gaseous and leached, which cause economic and environmental costs. Due to diminished N use efficiency stands out within the economic costs. The contribution of some N-based gaseous compounds that play a role in the greenhouse effect, acid rain, and the contamination and eutrophication of waters due to nitrate leaching further than the root zone are notable, among the environmental cost.

A series of changes in the management of arable land in the years 1967–1997 is being described in Alava, Attare and in this sense that progressively increased the concentration of N compounds in subsurface waters. For the human consumption by application of the European Directive 88/778/EEC related to water, and to nitrate pollution in the year 1999 the zone related to the Eastern sector of the quaternary aquifer of Vitoria was designated as vulnerable.

Where the wheat is grown in Alava this zone comprises of 38% of the area, 9500 ha approximately. Depending on the previous culture and soil richness the N fertilization is as much as 140 kg N ha−1 in this zone.

Nitrous oxide (N_2O) is not a very reactive gas, for as much as 150 years, it persists in the atmosphere. In various wavelengths in the infrared region between 7.7 and 17 μm and its greenhouse effect per mass unit is some 300 times larger than that of CO_2 this gas adsorbs electromagnetic radiation.

In the last 100 years, it is estimated that N_2O has contributed approximately 5% to the warming up of the planet in this sense. The origin of 90% of the N_2O emissions is anthropogenic, and agriculture is its main source. Due to both nitrification and denitrification N_2O in soils can be produced. In which ammonium first oxidizes to nitrite and then to nitrate is a microbial aerobic process of nitrification.

Figure 2.8. Nitrogen losses: gaseous and leached nitrogen balance.

Source: Image by the American society of agronomy.

N_2O can be released into the atmosphere, in this ammonium to nitrate oxidation process. And on the other hand, in which organic carbon is used as the energy source and the nitrate as the last electron acceptor so that it reduces to the last nitrogenous gaseous compounds N_2O and N_2 is a microbial anaerobic process of denitrification. In the soil, the nitrification and denitrification processes can occur simultaneously since the aerobic and anaerobic conditions can simultaneously take place at the same soil aggregate.

In a settled period of time at the soil-plant system, the N balance allows the knowledge of the N evolution. Other than the fertilization, it also enables the knowledge of the sources of N and the transfer between the different components of N, and the main mechanisms of N loss, and what amount of N is not likely to be recovered.

Therefore, in a culture system, the balance of N helps explain certain parameters that determine the dose of N needed by the cereal, optimizing the nitrogenous nutrition of the plant and reducing the danger of contamination.

To ascertain the quantity of N is the goal of this work, which is lost by leaching or by being released to the atmosphere, and to study the factors that have an effect on such losses. Along with this, in the soil-plant system was to be determined.

2.6. CHALLENGES TO SAFE WHEAT STORAGE

In the world wheat is one of the most important crops. About 2 million square kilometers it is cultivated on an area. To the several areas of the world, its ease of adaptation to the soil has been spread. For many countries in the world, wheat is the mainstay of nutrition.

The data was based on the latest report of the Food and Agriculture Organization of the United Nations (FAO), that the global wheat trade in 2017 reached 754.8 million tons; Russia's exports of wheat amounted to 30 million tons, America 26.5 million tons, Europe 25 million tons, Canada 21 million tons, Australia 20 million tons, while exports of Ukraine 15 million tons, Kazakhstan 8.5 million tons, Argentina 8 million tons, the largest countries exporting wheat worldwide

Fortunately, by an increase in global production, global wheat prices have fallen, which has been affected by good weather and the availability of water to grow in the EU, India, Pakistan, China, and the United States, which was accompanied by a rise in global exports.

In the year 2017–2018, it has been forecasted that the use of wheat will reach 740 million tons. By only the six countries more than 43% of the production is being consumed, such as China, India, Russian Federation, the United States of America, and Pakistan. This is because of the population of these countries of nearly 3 billion people, or half the world's population.

Figure 2.9. Challenges to safe wheat storage.

Source: Image by Pixabay.

There are many countries in the world who do not produce sufficient quantities of wheat for their uses and therefore tend to import this strategic commodity. There are some of storage places which are known as silos, which are being used by the wheat-importing countries to store large quantities of these grains. The wheat grains have been stored in huge shipments since the ancient times.

Grain stores has discovered earlier that these stores in the settlements of the Neolithic period before the pottery "A" were located in the Jordan Valley, where the first stores were located in places among other buildings. And at the beginning of 8500 BC, they were moved inside the house, but with the period of 7500 BC, they were stored in rooms dedicated to it.

From the outside, the area of the first wheat stores was 3 × 3 m, and in order to protect grain from rodents and insects, it had suspended layers and also provide ventilation. In the Mahjara, these stores are then located, since 6000 BC which is placed in the valley of Sindh. In the years of prosperity for use in drought ones the ancient Egyptians used to store wheat grains.

This is because the climate in Egypt is very drought, and without a significant loss of quality, Egyptians have been able to store grain in silos. In all the land of the East since time immemorial, an ideal way to store grains is called grain silo. Moorings used to buy wheat or barley in Turkey and Iran, which is relatively cheap and stored it in closed and hidden places in the face of famine.

In the hundreds of soils, relatively large quantities of wheat were stored and fug into rocks in Malta. By taking the proper precautions, the soils can store up to 60–80 tons, and keeping it in good condition for 4 years or more. For grain conservation stores specifically designed by the end of the nineteenth century, it began to spread in Great Britain, but North America was the home for the major stores, called grain levers.

To grain storage in Great Britain on those difficulties significantly there were large-scale climatic obstacles. It has been said that grain should be kept as far away as possible from moisture and heat in order to keep the grain in good condition because new grains tend to release moisture when brought to the store.

On the mushroom-shaped logs, small wheat stores were built, which is known as the stone pillars in Great Britain. On a wooden frame, it was built and usually has stone roofs. From the front, the large one resembles the open ceiling, but the upper part is closed. By a stone staircase on the outer wall, the first floor is usually accessed.

In the construction of silos with tremendous advances in engineering technology, a new generation of forms and possibilities of silos has begun. In farms, mills, and harbors high density indoor silos are found and can reach a height of 30 m.

In North America largest silos are located at major shipping centers such as Chicago, Kansas City, and Missouri in the United States, and Thunder Bay, Ontario, and Canada. Many cylindrical silos are equipped with a perforated metal floor that allows the air to pass through to keep grain-free from moisture. With high mechanical devices nearby, cylindrical silos are often adjacent.

Basically, there are three types of silos, the first type is made of wood, the second type is made of concrete, and the third type is made of metal (steel and galvanized). Wheat is as barley, oat, rye, and so on, used to store suitably both in concrete and metal silos.

Including dryers, detergents, workpieces, cranes, and conveyors, large silos have many appliances. Dryers leave grain which is free from moisture. Over the grains, the air is heated in the dehydrator and directed, and then it is used in the air that has not been heated to cool the grains.

Grain cleanliness is being ensured by the use of detergent. To the elimination of grains, the absorption of dust, husk, or straw is obtained and screening and vibration leads that are not of the appropriate size and density. However, to isolate rotten grains, some machines use photovoltaic cells.

To transport the grains, cranes and conveyors are being used. By cranes vertical grain movement is being obtained. With a large conveyor belt and a bundle of buckets among the most used cranes is a crane. To the high-altitude wheat grains in the bucket are transported and poured into the storage baskets. Across the silos tankers move the grains horizontally, which is called a grain lift. It is a silage storage place as well.

To the silos the large quantities of wheat are being delivered. Of course, from the farms, the huge quantities received by silos are coming directly without any transactions. With all the residues of process of flail, agricultural soil wheat is then loaded and contaminants of harvesting tools during the separation of grains from the ears. To prepare wheat grains for the flour process, it is perhaps of standard importance to explain how grinding of wheat grains is done by the following procedures.

Of the three main parts, the grain of wheat consists of, grain coats, embryos, and endosperm. To separate as much of the endosperm content as possible from the wrappers is the purpose of the milling process. In the production, the ratio of flour produced to the percentage of grain used is known as the extraction.

From the 100 g of grain, the weight of flour is being produced, and it depends on this percentage on several factors, the most important; extraction method used, type of mills used, nature, and specific weight of wheat grains. In general, extraction rate is between 70 and 72% in excellent white flour, and this percentage is 90–95% in brown flour.

2.6.1. Receiving

From the fields or silos, the grain which is coming is received in the mill after sampling and examined to ensure that it meets specifications set by the mill. In the conical tank's grains are received and emptied, each of which can reach 10 tons. With a fixed metal mesh for impurities when unloading grain of which it is being covered. For delivery to the cleaning equipment, the crane and aspirator pull the grain out of the tanks.

2.6.2. Cleaning

The two main units is being consist by the cleaning equipment: dry-cleaning unit known as the black cleaning unit and the wet-cleaning unit known as the white cleaning unit.

2.6.3. White cleaning

This unit consists of a peeler and stalker.

With the rough internal surfaces or coated with a precision metal bed the peeler consists of a cylinder that scrapes centrifugal grains. By an air stream generated shells are being separated and by an electric fan installed at the top of the peeler.

With a nozzle to feed it with the grains, the asset consists of a rectangular basin, as well as an appropriate water source that can be controlled as needed, and ends with a drainage basin topped by an appropriate filter. In the opposite direction inside cylinder with two helical separators working that wash the grains and push them forward and the light impurities on the surface of the water. It passes over a vibrator strainer that has been removed from the water and then to a rotary roller dryer to complete removing water droplets from the grain, with the rush of grain forward.

2.7. QUALITY ASSESSMENT OF FEED WHEAT IN RUMINANT DIETS

As a feed grain, wheat is not used traditionally because its milling properties make it desirable for use in breads, pastas, and noodles; the milling of wheat produces flour for human use and appreciable quantities of by-products for animal feeds. However, in ruminant feeding, there have been significantly increased supplies of feed wheat used as livestock feed.

In increasing the wheat portion in animal rations, the livestock industry is interested because of competitively priced wheat compared with other feed grains. Wheat is low in fiber and high in starch and protein. In the form of carbohydrates feed wheat is primarily used as a source of energy, which is similar to other cereal grains. As the digestible energy or metabolizable energy per unit of dry matter is similar to corn the way in which available energy is being expressed but higher than other major grains.

For poultry, the majority of wheat has traditionally been used as feed and swine because the use of large quantities of feed wheat in ruminant rations has a number of concerns and problems. Wheat starch is highly fermentable and which is the problem that appears, and its rate of digestion in the rumen is greater than that of corn and barley grains that increase the risk of rumen acidosis in animals fed high-grain diet.

Moreover, due to different types of wheat: soft, hard, and durum, and the growing conditions, the physical characteristics and nutrient content of wheat can vary considerably. Therefore, when quality of feed wheat is assessed nutrient contents of wheat, variety, and growing conditions need to be considered. In addition to it, factor affecting wheat quality is another key of kernel processing.

Figure 2.10. Quality assessment of feed wheat in ruminant diets.

Due to the resistance of the seed coat to attack by rumen microbial or host enzymes, whole wheat kernels are poorly being digested. Though, to break seed coat wheat grains require processing. From about 60% if fed whole to over 90% if properly processed the digestibility of wheat can be increased. However, for best results when feeding wheat since over-processing can result in digestive upset and several factors (kernel hardness and uniformity, and processing methods) can significantly affect the processing results, attention to processing is crucial.

Finally, in wheat grain level of deoxynivalenol may affect the feed value. As the vomitoxin the deoxynivalenol, commonly referred and it is mycotoxin that may be produced in wheat and barley grain infected by Fusarium head blight or scab.

2.8. PHYSICAL AND CHEMICAL CHARACTERISTICS OF FEED WHEAT

It depends upon the end-use; wheat quality is a complex term. To the energy level, the quality should be associated primarily for the feed wheat and also the protein content as well as its digestibility in the digestive tract of animals. With its nutrient content such as content of starch or protein, the feed wheat quality can be assessed and physical characteristics like thousand kernel weight, test weight, and kernel hardness which are easily measured and commonly used by commercial feedlots and feed mills to assess the quality of wheat as animal feed.

Figure 2.11. Physical and chemical characteristics of feed wheat.

Source: Image by Wikimedia commons.

By planting season, hardness of the grain and color wheat cultivars can be classified. In the fall-winter wheats are winter hardy, so they are planted. In the summer wheat harvested early and, in the spring, they resume maturation. Spring wheats are planted in the spring and harvested late in the summer. Soft wheat varieties have starchy kernels and less gluten which mill easier than the hard wheats, whereas hard wheats have higher protein and gluten levels than the soft wheat with the hardest wheat is durum.

2.9. CONCLUSION

To conclude, wheat comes as one of the essential crops, and its production is being processed with a variety of factors such as climatic factors and its usage and growth with different chemicals. Wheat comes with very complex growth system which requires proper cleaning and storage facilities.

Moreover, the consumption of wheat and its nutrients depends highly on the environment and season it is being grown, having different effects on diet plans on overall health. Furthermore, apart from challenges, wheat comes as a high demand for consumption globally and its production needs to be operated in a decisive manner.

REFERENCES

1. Ahmed, M., Khan, S., Irfan, M., Aslam, M., Shabbir, G., & Ahmad, S., (2018). Effect of phosphorus on root signaling of wheat under different water regimes. *Global Wheat Production*. [online] Available at: https://www.intechopen.com/books/global-wheat-production/effect-of-phosphorus-on-root-signaling-of-wheat-under-different-water-regimes (accessed on 7 July 2021).
2. Iqbal, M., (2018). Role of osmolytes and antioxidant enzymes for drought tolerance in wheat. *Global Wheat Production*. [online] Available at: https://www.intechopen.com/books/global-wheat-production/role-of-osmolytes-and-antioxidant-enzymes-for-drought-tolerance-in-wheat (accessed on 7 July 2021).
3. Moustafa, S., Alhaithloul, H., & Abdelzaher, H., (2018). Challenges to safe wheat storage. *Global Wheat Production*. [online] Available at: https://www.intechopen.com/books/global-wheat-production/challenges-to-safe-wheat-storage (accessed on 7 July 2021).
4. Ortuzar-Iragorri, A., Castellón, A., Besga, G., Aizpurua, A., Fuertes-Mendizabal, T., & Estavillo, J., (2018). Nitrogen losses: Gaseous and leached nitrogen balance. *Global Wheat Production*. [online] Available at: https://www.intechopen.com/books/global-wheat-production/nitrogen-losses-gaseous-and-leached-nitrogen-balance (accessed on 7 July 2021).
5. Szilvia, V., Gabrijel, O., & László, Z., (2018). *Wheat Sensitivity to Nitrogen Supply under Different Climatic Conditions*. [online] Intech Open. Available at: https://www.intechopen.com/books/global-wheat-production/wheat-sensitivity-to-nitrogen-supply-under-different-climatic-conditions (accessed on 7 July 2021).
6. Yang, W., & Shen, Y., (2018). Quality assessment of feed wheat in ruminant diets. *Global Wheat Production*. [online] Available at: https://www.intechopen.com/books/global-wheat-production/quality-assessment-of-feed-wheat-in-ruminant-diets (accessed on 7 July 2021).

Chapter 3

Impact of Climate Change on Grassland

CONTENTS

The chapter on impacts of the climate on grassland explains the basic significance of the impact of climate change on grassland production and soil carbon all across the globe, which includes grasslands and the global carbon cycle, and evaluation of the grassland response to climate change. This chapter also explains the impacts of the increased atmospheric carbon dioxide concentration and carbon monoxide climatic interaction such as detecting changes in the grasslands.

This chapter also explains the effects of the climate change on grassland biodiversity and productivity, which emphasize a summary of model types. This chapter addresses the impact of the climate changes on the grassland such as impact of the climate change, impact of the climate change and atmospheric carbon dioxide concentration on species composition, and impact of climate change and atmospheric carbon dioxide concentration on grassland nutrition.

This chapter also mentions the shifting plant species composition in response to the climate change stabilizes grassland primary production. This chapter also includes the climate change effects on forests and grasslands such as benefits/stresses, effects, water/precipitation, water/snow cover, water/flow, water/sea-level rise, and vegetation/growing season.

3.1. INTRODUCTION

Sudden changes in the climate impacts with respect to the grasslands and prairie bioregions is consisting of increment in the seasonal, annual, minimum and maximum temperature and altering perception patterns. Due to these ecosystems are comparatively dry with a strong seasonal climate, they are very sensitive with respect to the climatic alterations and vulnerable as well in the context of changes in the climate regime.

For instance, model stimulations depict that the regional drought in the Prairie Pothole Region may result in the loss of precious habitat for breeding the water flow, from a region that historically has harvested near about 50% to 80% of ducks of the continent. Hotter as well as dryer circumstances may also takes into the direction towards the violation of the new species, and a higher chance of risk of wildfire.

Increment in the temperatures, lesser rainfall, and drought are already being noticed in several numbers of regions, and the arid especially in South-west is projected in order to become even drier in this century. The forest region that are wetter are likely to intrude the existing savannas, on the other hand, increment in the arid region deserts projected in order to expand to some level and move upward in elevation, as an outcome this produces "desertification" of the arid grassland ecosystems.

One of the most major aspect that impacts the grassland and prairie ecosystems is drought. Drought is likely to be led to the increment in the wildfire as well as loss of wetland habitats, for example, prairie potholes that are very important habitat with respect to the migratory bird species, along with the species migration and habitat shifts.

Figure 3.1. Basic significance of the climate change on grassland.

Source: Image by Pxhere.

Vegetations shifts from C3 to C4, the communities of grassland and phenological changes will influence or impact the ecosystem as well as species, and the modification in the composition of species along with the productivity of the plant that may also influence or affect the human communities that depends on the agricultural production in these kinds of areas. According to the NASA satellite imagery shows that the anticipated modifications in the plant productivity are already take place all across the globe.

Slight modification in the temperature and rainfall can changes the composition, abundance, and distribution of species in arid lands in a considerable amount, and the products as well as services they offer.

For instance, observed and projected decrement in the regularity of freezing temperatures, lengthening of the frost-free season, and increment in the minimum temperatures can change the species of plants and the alteration of the geographic and elevational boundaries of several numbers of arid lands.

The level of these alterations or modification will also rely on the shifts in the precipitation and fire. Increment in the frequency of the draught could also produce major changes with respect to the vegetation. Losses of vegetation cover that are integrated with the increases in the intensity of the precipitation and climate-induced reductions in the soil aggregate the stability will enormously increase the rates of erosion.

Transport of eroded sediment to streams integrated with the alterations in the timing and the amplitude of the minimum as well as maximum flows can impact the quality of water, aquatic fauna, and riparian vegetation. Land trusts and the conservation agencies are already observing these kinds of shifts and are working in order to identify the risks, decrement in vulnerabilities, and regulate the grassland systems for changing conditions.

3.2. IMPACT OF CLIMATE CHANGE ON GRASSLAND PRODUCTION AND SOIL CARBON WORLDWIDE

The effect of the sudden changes in the climate and the increment in the atmospheric CO_2 was modeled for 31 temperate and tropical grassland sites, with the help of the CENTURY Model. Sudden changes in climate elevated the net primary production, except in cold desert steppe regions, and CO_2 increased production everywhere. Sudden changes in the climate produce soil carbon to decrease overall, with a loss of 4Pg from global grasslands after the time period of five decades.

Both sudden changes in climate and increased $CO_{2,}$ altogether, results in increment in the production and decrement in the global grassland carbon losses to 2Pg, with tropical, savannas becoming small sinks for soil C.

Determination of statistically significant modification in the production of plants would require a change of near about 16%, which is measured in the production of plant because of high year-to-year variability in the plant production. Most of the anticipated alterations in the plant production are less than 10%.

3.2.1. Grasslands and the Global Carbon Cycle

The potential impacts of the sudden changes in climate on the grassland biochemistry have received much less attention as compared with the forests (Hall & Scurlock 1991). The impacts of the modifications or alterations in the temperature, water as well as nutrients are comparatively well acknowledged, and the interactions with the long-term impacts of the carbon dioxide fertilization are now starting to be sufficiently well-developed in order to calculate the alterations in the carbon fluxes (Schimel et al., 1990, 1991; Long 1991; Owensby et al., 1993).

Figure 3.2. Grasslands and the global carbon cycle.

Source: Image by Pxhere.

Grassland ecosystems store most of their carbon (C) inside the soils, where turnover times are comparatively long (hundred years to ten thousand years), and so alterations, nevertheless, they can take place slowly, will be of considerable interval of time. Alterations in the grassland carbon storage will have a significant as well as long-lived impacts on the carbon cycles all across the globe.

Fluxes of carbon with the help of plant and soil organic matter in grasslands, more particularly in the tropical regions, may have been underestimated in the plants (Long et al., 1989). Grasslands are one of the most extensive types of vegetation all across the globe, which covers near about 20% of the land surface of the world's land, that is (24 x 106 km^2).

Grasslands are consisting of more than 10% of the soil carbon stocks all across the globe (Eswaran et al., 1993). Grasslands are also a significant or meaningful habitat with respect to the wildlife, humans as well as their domestic livestock and are subject to conversion to both irrigated agriculture as well as dryland.

Under the guidance of the integrated SCOPE (Scientific Committee on Problems of the Environment) project (Breymeyer & Melillo 1991), the CENTURY model of the interactions of the soil and plant has been used to an extensive range of grassland all across the globe (Parton et at. 1993).

The SCOPE- GRAM group has accumulated or collected long-term data from near about 30-one sites in the context of both tropical grassland as well as temperature. These collected data are then used to evaluate the robustness of the CENTURY model and the simulated response of the grasslands with respect to the scenes that are related to the sudden changes in the climate.

Various numbers of recent studies have approximated the response of the soil organic matter (SOM) in the context of environmental change. Jenkinson et al. (1991) approximated the considerable as well as quick emission of carbon from soils, which, as an outcome produces global warming. On the other hand, Schlesinger (1990) debated that the undisturbed soils will have a low potential for added carbon storage and can only changes with a slow rate.

According to most of the previous studies, that have had recognized the limitations or disadvantages. Among from that, studies done by the named researcher Schlesinger, who used inferential soil chrono-sequence studies. On the other hand, the model of Jenkinson et al. (1991) did not well think out the feedback between the soil processes and the sudden changes in the climate impacts the primary production, which is a prime aspect of CENTURY.

Certainly, Jenkinson et, al. (1991) observed that, in order to find out how the sudden changes in climate are impacting the flux of carbon dioxide (CO_2) between the soil and the atmosphere, their model 'would have to be integrated with the another which specifies how plant production all across the globe is changed with the help of increment in the concentrations of the greenhouse gases (Jenkinson et al., 1991).

3.2.2. Evaluation of Grassland Response to Climate Change

Sites were selected in order to represent an extensive range of the regions of the grassland, that are based on the climate factors as well as properties of the ecosystem. This is consisting of near about 15 temperatures and the tropical regions for which a detailed data on a monthly basis were available on the vegetation biomass and the plant debris. A detailed comparison of the stimulated and observed peak live biomass as well as the plant production for these 15 regions, that depicts a good correlation (N = 0.45 and 0.70, respectively).

Soil carbon level were anticipated very well (N=0.93). Researchers have assembled at least 20-five years of weather data on a monthly basis for 15 various sites and a further 16 grassland sites that are derived from the literature. The sites are ranged from temperature ecosystems with patterns of highly seasonal temperature to the tropical areas where the temperature fluctuates very little. Precipitation at all of these regions showed a marked seasonality, and in most of the scenarios, a high level of variability has been shown in the time period of several years.

With respect to the temperature, seasonality of the climates, and rainfall, there are a total of 30-one sites into seven bioclimatic areas that are based on a modification of Bailey ecoregions, that are well thought out as ecological factors as well as climatic factors (Bailey 1989) (as shown in the Tables 3.1. and 3.2). The three divisions of Bailey's dry domain are based on the temperatures, with the cold desert steppes that are having regional mean annual temperature of near about −0.3°C. The temperate steppes having near around 10.2°C and the dry savannas are near around 24.1°C.

Table 3.1. Long-Term Mean Annual Precipitation, Temperature, Potential Evapotranspiration Rate, and Ratio of Annual Precipitation to Potential Evapotranspiration Rate (PPT/PET), for the Seven Grassland Regions Based on Bailey (1989)

Biome type	Mean annual precipitation (mm)	Mean annual temperature (°C)	Mean annual potential evapotranspiration	PPT/PET	Global area ($10^6/2$)	Vegetation type
Cold Desert Steppe	299	–0.3	960	0.334	2.095	C3
Temperate Steppe	301	10.2	1041	0.294	2.943	C3/C4
Humid Temperate	700	10.2	1041	0.294	2.943	C4
Mediterranean	497	15.7	1253	0.401	0.161	C3 C4
Dry Savanna	387	24.1	1977	0.165	5.106	C4
Savanna	788	23.4	1491	0.554	7.990	C4
Humid Savanna	1555	27.4	1404	1.195	1.708	C4

Source: Table from Reseacrchgate.net.

Table 3.2. Simulated Mean Annual Above- and Belowground Plant Production, Total SOC, mean Annual Abiotic Decomposition Factors, and Net N Mineralization Rates for Each of the Seven Grassland Regions Prior to Climatic Changes

Biome type	Aboveground production (gm^{-2})	Belowground production (8*) 2	Total Soil C (0–20 cm) (gm^{-2})	Net N mineralization (g N m* 2y*')	Abiotic decomposition rate (0–1)
Cold Desert Steppe	60	92	5530	3.11	0.093
Temperate Steppe	36	59	2020	2.67	0.158
Humid Temperate	134	161	7171	4.81	0.194
Mediterranean	79	103	3005	3.44	0.184
Dry Savanna	55	85	2275	2.58	0.173
Savanna	191	222	4306	5.73	0.339
Humid Savanna	340	344	3273	6.61	0.644

Source: Table from Researchgate.net.

3.2.3. Model Runs Under Climate Change Conditions

The doubled carbon climatology that is used in order to drive the CENTURY model in the present study were derived from two Global Climate Models (GCMs), the Canadian Climate Centre (CCC) and the Geophysical Dynamics Laboratory High Scenario (GFHI). These models are used to offer assistance as they provide monthly projections of the temperature and the precipitation for 2.5 degrees grid cells all across the globe.

The approximated values for a doubling at atmospheric carbon concentration, and assumed that these would be reached within the time interval of near about five decades from the present interval of time, corresponding roughly to IPCC Scenario A (Houghton et al., 1990). Anticipating alterations with respect to the annual as well as seasonal temperature and precipitation.

The two global climate models simulate identical alterations with respect to the annual air temperature (2–5°C). Nevertheless, the alterations in the precipitation were different between the models. The Geophysical Dynamics Laboratory High Scenario (GFHI) model depicted the increment in the precipitation for all of the areas. On the other hand, the Canadian Climate Centre (CCC) model shows the decrement in the precipitation for the temperature steppe and humid savanna regions and the increment for the other regions.

3.3. EFFECT OF INCREASED ATMOSPHERIC C CONCENTRATION AND CO_2 CLIMATE INTERACTION

The effects of the increment in the atmospheric carbon concentration on the grassland ecosystems with and without the inclusion of the sudden changes in the climate. Overall, the impact of the increment in the atmospheric carbon concentration under the present patterns of conditions of climate was to increase the total plant production, the rate of abiotic decomposition and the SOC storage. On the other hand, nitrogen mineralization rated falls.

Production elevated the most of in the cold desert steppe region, and SOC storage elevated most for the humid savannas. Soil carbon increases were negatively correlated with the alterations in the rates of the decomposition. Nitrogen mineralization rates are fallen in several regions, as an outcome from an elevated input of the lower quality litter with the higher carbon to nitrogen ratio (Melillo et al., 1982), but elevated for the cold desert steppe region. The stimulated elevates in the decomposition under the increased carbon, resulted from the enhanced soil water relationships at the cold dry sites (cold desert steppe, temperature steppe), with the smaller increment for the warmer and more humid regions. When the relative alterations in the plant production, SOC, nitrogen mineralization, and the decomposition were compared between

1. the integrated effects of the elevated atmospheric carbon (C) concentration and climate; and
2. the independent effects or influences of carbon and climate change, a linear additive effect was demonstrated (N = 0.97, 0.98, 0.97 and 0.91, respectively).

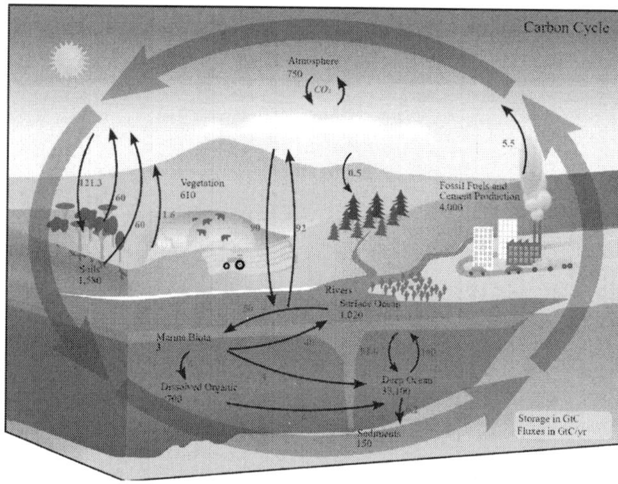

Figure 3.3. Effect of increased atmospheric c_2 concentration and co climate interaction.

Source: Image by Flickr.

Elevated atmospheric carbon C_2 concentration helps in improving the total plant production regardless of the impacts of the sudden changes in climate, consistent with the results of Melillo et al. (1993).

This improvement in plant production helps in reducing the SOC losses which as an outcome from climatically driven alterations in SOM decomposition. The direct carbon C_2 impacts the decrement in carbon losses all across the grassland systems and results in a SOC sink region in the tropical savanna and humid savanna areas, irrespective of climatic impacts.

The effects of these climate and carbon alters the present grassland SOC stocks shows that the climatic effects alone will produce a net loss of near about 4 Pg. carbon (as shown in Table 3.3). With all regions becoming carbon sources as an outcome of decomposition of SOC. Nevertheless, the effect of the increment in the atmospheric carbon C_2 is to offset the SOC losses due to the sudden changes in climate.

The effect of carbon decreases the net losses all across the globe from grasslands by near about 50% in the Canadian Climate Centre case and 66% in the Geophysical Dynamics Laboratory High Scenario case, recommending a net carbon source of 2 Pg. C, over the time period of five decades, or 0.04 Pg. C per annum.

Talking about the tropical regions, savannas along with the humid savannas actually become SOC sinks with the inclusion of Carbon C_2 and the sudden changes in the climate altogether, irrespective of the condition of GCM.

These outcomes fluctuate with those of Jenkison et al. (1991), who approximated a net source of 61 Pg. C, all across the globe, over the time period of next six decades from the soil organic matter, all across the world.

Grassland soils (such as temperate steppes and tropical savannas) that accounted for near about 20% of the world SOC stocks in this analysis, recommending a grassland source of near about 12 Pg. C, over the time period of six decades, or 0.2 Pg. C per annum, that is based on calculations of Jenkinson.

3.3.1. Detecting Changes in Grasslands

The potential to determine the effect of the sudden changes in climate with respect to the production of plant and SOC levels, given the observed natural variability, that was assessed with the help of CENTURY model results.

The 20-five average plant production and the levels of SOC for all of the 30-one sites that are prior to the sudden changes in climate were compared with the average values for the period of 20-five years following the start of the changes in climate for both of the GCM situations. Along with that, a statistical f-test was used in order to find out, if the mean values were considerably different at the 5% level.

Near about 8% of the 62 model runs showed the considerably different plant production (predominantly those with the low productivity), with the alterations in the plant production averaging +22% (that ranges from 17% to 31%). The nonsignificant comparisons had an average shift in the plant production of +7% (that ranges from zero percent to 16%).

Around 81% of the comparisons depicts the considerable shifts in the mean SOC, with SOC changes averaging 7% (that ranges from 1% to near about −33%). SOC changes of less than 1% were nonsignificant.

These outcomes recommend that the stimulated SOC and the plant production would have to shift by the value of at least 1% and 16%, in order to be statistically detectable. Noting that it would be impossible to determine 1% change in SOC in the field because of the large spatial variability in SOC (Yonker et al. 1988), and it would take near about 5% to 15% to be detectable.

According to the several numbers of comparison that shows considerable amount of fluctuations or variations with respect to the plant production and SOC did not change with time (for the time span of 25–50, 50–75, and 75–100, following the start of the climate change).

Modifications in the plant production or SOC that is induced with the help of sudden changes in climate and increment in the content of carbon present in the atmosphere will therefore have to be considerable, overcoming interannual variability and sampling variance, respectively, in order to be detectable.

Comparison of the anticipated alterations in the plant production with this statistical analysis recommends that most of the anticipated shifts in the plant production would be very hard to find out. Nevertheless, the plant production has stabilized 70-five years after the start of the climate change, SOM has not reached a new steady-state, even though the rate of alteration in the SOM has been reduced.

Comparison of the transient and long-term (near about 5000 years) equilibrium levels of SOM depicts that only 26% of the long-term shifts in SOM has took place near about 50 years to 70-five years after the initiation of the alterations in climate. In this way, it is very important to differentiate among the transient responses that are induced instantly with the help of sudden changes in climate and the long-term equilibrium response.

According to Bailey's humid tropical domain that was divided into the savannas and humid savannas on the basis of precipitation, the former receiving a regional annual mean of near about eight-hundred-millimeter and after that it become near about 1550 millimeters. The stimulated outcomes for each bioclimatic region were assessed with the help of averaging results from the regions that are in a particular bioclimatic zone.

3.4. EFFECTS OF CLIMATE CHANGE ON GRASSLAND BIODIVERSITY AND PRODUCTIVITY

There is an increment in the proof that the influence of the sudden changes in the climate with respect to the productivity of the grasslands will at least partly relies on their biodiversity. A high level of biodiversity can discuss the stability to the grassland ecosystems against the environmental change, but there are also direct impacts of the biodiversity on the quantity as well as quality of the productivity of the grassland.

Figure 3.4. Effects of climate change on grassland biodiversity and productivity.

Source: Image by Pixabay.

In order to explain the manifold, the interactions, and in order to anticipate the future climatic responses, models can be used. Nevertheless, model designed for the study purpose, the interaction between the biodiversity and the productivity tends to be structurally different from the models for studying the climatic effects.

Permanent grasslands are every so often hot spots of the biodiversity, which take participations to the temporal stability of their services. The variety of the plant species that are present in the grasslands is strongly impacted by the long-term management practices, with the livestock grazing being the major driving force that are impacting the vegetation dynamics, species distribution and the landscape-scale biodiversity in addition to forage quantity and quality.

Biodiversity, which is surrounded by the several types of species and across landscapes, can be crucial for the longer-term resilience of the ecosystem functions and the services that they underpin. Biodiversity is closely connected to the ecosystem services with the help of several numbers of different relationships, with the communities that are full of species are tending to perform better as compared to any other individual species.

Biodiversity-ecosystem functioning relationships are impacted with the help of several numbers and identities of the species, their evenness within the community, their functional characteristics, and their interactions. If the species loss can be compensated with the help of other species that are taking participation similarly to functioning, the role of the individual species may shift with the environmental change.

Then, abiotic alterations can lead to the biotic alterations and vice versa, and the various kinds of species that take participation to most to any given ecosystem process at different points in the time and space. Nevertheless, there are limits to the species redundancy, and high biodiversity is required in order to maintain the several kinds of processes that are operating in multifunctioning ecosystems.

As concepts of the ecosystems functioning have evolved, work has broadened to surround the biodiversity loss within and between the trophic levels. This work has offered several advantages from the studies of food webs and more extensively of ecological networks.

Nevertheless, according to the researcher named as Novak, who concluded that these kinds of studies can only offer a limited predictive capacity. On the other hand, the general knowledge of the strengths of interactions between several kinds of species remains poor.

The requirement for the elevated predictive capacity is pressing, as the ecosystem of the world undergo the unprecedented modifications with the species being lost from an extensive range of the ecosystems and the trophic levels and there is a requirement in order to well thought out the role that models can play.

Particularly speaking, the challenges that grassland systems are coping up in the present interval of time, which shows that the aspects that are related to the diversity of the plant that cannot be ignored in modeling studies. In this way, grassland models are required to consider the nutritional value of the multispecies swards.

The differing herbage intake of grazing animals between mixtures and the pure stands, the dependence of the content of the milk protein on the botanical composition of swards, the beneficial effect of legumes on the nitrogen economy of the multispecies swards, and the supporting and controlling services (e.g., pollination, drought resistance, and pest control) provided with the help of multispecies swards.

With respect to the primary importance, the requirement in order to account the complicated and variable relationship between the grassland biodiversity and the productivity. If primary productivity is not used with the help of grazers or humans, then it may sustain the larger biodiversity, nevertheless, some level of grazing may stimulate biodiversity. In addition to this, decreasing the intensity of the management (e.g., fertilization) in grasslands decreases the livestock productivity but may improve the quality of the meat.

In this way, trade-offs between the diversity and the way grassland productivity are exploited exist, and sudden changes in the climate can impact how to prioritize one or the other. Modeling tools are required in order to negotiate these trade-offs.

The literature on the grassland modeling is extensive. Google scholars find more than 2 million articles that are consisting of the terms such as 'model' as well as 'grass' or 'grassland,' and 1670 of those have the terms in their title. Web of Science finds 1403, in this way, titles papers (information retrieved on 20 November 2017).

The primary interest here is in modeling the responses of grasslands of different kinds of levels of biodiversity with respect to the sudden changes in the climate. Of course, the climate not only impacts the grassland biodiversity and the productivity but is itself impacted with the help of dynamics at large spatial scales.

Grasslands, such as all vegetation, impacts the climate with the help of their albedo and greenhouse gas balance, and also can play a vital role in mitigating the climate change.

The impacts of the vegetation on the atmosphere are depicted in the latest generation of GCMs (Global Climate Models), but are not further discussed in this chapter.

A proper consideration is required for both static as well as dynamic modeling, and modeling is objected at explaining the observations along with the modeling is objected at predicting the influence of environmental change.

After a brief review of empirical models, process-based models, and integrated grassland modeling methodologies. These are reviewed separately, before discussing their relative strengths and weaknesses, and the scope for using elements from one model type in another. A summary of this model overview is provided in Table 3.3.

Table 3.3. Summary of Model Types

	Empirical Models	Process-Based Models			Integrated Models
		Ecological	Biogeochemical	Agricultural	
Characteristics	Statistical models (regression, etc.)	Small models focusing on biotic interactions; elegant mathematics	Parameter-rich models of cycling of carbon, nutrients, water	Parameter-rich models of phenology and yield formation	Hierarchical models that include process-based models and human agents
Static/Dynamic	Static	Dynamic	Dynamic	Dynamic	Dynamic
Strengths	Flexibility; ease of use; spatial heterogeneity	Explaining general ecological patterns including species interactions	Long-term prediction of biogeochemical cycles and soil pools	Short-term prediction of productivity	Comprehensive; scope for assessing policy
Weaknesses	Non-predictive outside calibration range	Abiotic effects and spatial heterogeneity poorly accounted for	Limited capacity for yield forecasting; no spatial heterogeneity/biodiversity	No long-term predictive capacity; no spatial heterogeneity/biodiversity	Weaknesses inherited from component models; difficult to parameterize and test

Source: Table from mdpi.

3.5. IMPACT OF CLIMATE CHANGE ON GRASSLAND

Grassland accounts for 70% of the UK agricultural land and approximately 32% of agricultural output (IGER, 2003). Increased concentrations of atmospheric CO_2 and changes in climate in the future are likely to have an impact on grassland and production of grazing animals. Effects might include changes in the grazing season, yield, species and nutrient composition.

3.5.1. Impact of Climate Change

The UKCP09 future weather projections suggest overall warming accompanied by wetter winters and dryer summers (Jenkins et al., 2009). Grassland is sensitive to drought (MAFF, 2000) and dryer summers in the future might place limitations on grass growth on soils with low soil available water in the summer months (St. Claire et al., 2009).

More drought resistant species might be required to meet livestock grazing needs, or increased costs associated with additional feeding might occur. Drought will also increase the risk of wildfires affecting extensive grazing areas.

Figure 3.5. Impact of climate change.

Source: Image by Flickr.

Wetter winters may cause waterlogging and problems of poaching, possibly resulting in animal having to be housed, and reduced opportunities for grazing and harvesting (Hopkins and Del Prado, 2007). Good soil structure to promote resilience to the effects of drought and waterlogging should be encouraged (IGER, 2004).

Draining waterlogged soil would encourage early grass growth, but at the expense of a reduction in water availability in July and August (Armstrong and Castle, 1992).

Temperature rise is likely to increase the length of the grazing season (Topp and Doyle, 1996a; IGER, 2003) with grass production starting earlier in the spring and continuing later into the autumn. This could have advantages for the availability of grass during early lambing.

Although the rate of photorespiration increases with temperature (Nowak et al., 2004), the potential for increased herbage will be limited by nitrogen availability and, in some areas, irrigation during dry months and drainage during wet months (IGER, 2004).

The longer growing season may result in increased grassland yield (ADAS, 1997) though other models suggest that the longer growing season might only increase clover production and have little effect on yield overall after the impact of drought is taken into account (Topp and Doyle, 1996a; Grime et al., 2008).

3.5.2. Impact of Climate Change and Atmospheric CO_2 Concentrations on Species Composition

Alterations in the temperature, alteration in rainfall, and atmospheric concentration of carbon dioxide (CO_2) might impact the different forage species in several numbers of different ways, in this way, changing the composition of the grassland community. Overall conditions are likely to become more favorable for forage legumes (Soussana and Luscher, 2007).

Management factors like frequency of defoliation will also be important in finding out that how the species diverse communities respond to alteration in climate and atmospheric carbon dioxide concentration (Teyssonneyre et al., 2002; Harmens et al., 2004).

C3- Grasses, that have higher content of protein as compared to the C4 grasses and in this way, they have higher nutritional value, and will help in order to elevate the competitive advantage over the C4 grasses under the conditions of elevated atmospheric carbon dioxide concentration (Ehleringer et al., 2002; Soussana and Luscher, 2007).

Nevertheless, the anticipated increment in the temperature will favor the C4 grassland species (Soussana and Luscher, 2007), in this way, the outcome remains difficult in order to anticipate. Generally speaking, the grasslands that are rich in species are likely to be more productive with respect to the increased levels of the atmospheric carbon dioxide concentration as compared to the species-poor grasslands (Reich et al., 2001).

3.5.3. Impact of Climate Change and Atmospheric CO_2 Concentrations on Grassland Nutrition

The prime factors that find out the nutritional quality of grassland are; dry matter content, digestibility, energy and protein (HCC, 2008).

The protein provided with the help of forage is associated to the content of the nitrogen (Rinehart, 2008) and increased atmospheric carbon dioxide concentration have been shown to be linked with the decreased shoot nitrogen (Soussana et al., 1996; Ehleringer et al., 2002; IGER, 2003), along with the condensed vitamins (Soussana et al., 1996) and elevated soluble carbohydrate (Ehleringer et al., 2002; IGER, 2003).

The age of the plant also impacts the nutritional content of forage. Talking about the older plants, older plants have less protein, more fiber, and lower digestibility (National Research Council, 1985; Rinehart, 2008).

In this way, increment in the rate of growth are linked with elevated atmospheric carbon dioxide concentration, which might permit the younger plants to be consumed, and in addition to that, it might help in order to elevate the protein as well as the soluble carbohydrates content for livestock (Rinehart, 2008).

Nevertheless, the exact effect of the sudden changes in the climate on grassland nutrient content remains unclear, there are several numbers of indications that decreases the protein in grazing might influence negatively on weight gain in cattle (Owensby et al., 1996 in (Topp and Doyle, 1996b; Ehleringer et al., 2002; Craine et al., 2009). According to the several numbers of published studies about the impact of climate change on nutrition of grazing sheep appear lacking to date.

As stated by (Campbell et al., 2000), in the coming interval of time, the quality of the forage could decline in systems where the feed conversion productivity is limited by protein and can elevate or remain same in the systems where it is limited with the help of availability of the carbohydrate.

3.6. SHIFTING PLANT SPECIES COMPOSITION IN RESPONSE TO CLIMATE CHANGE STABILIZES GRASSLAND PRIMARY PRODUCTION

The structure as well as the functions of the alpine grassland ecosystems, which is consisting of their broad soil carbon stocks, are largely shaped with the help of the temperature. In particular, the Tibetan Plateau has experienced a considerable warming over the time interval of around five decades, and this warming trend is anticipated to increase in the coming interval of time.

These kinds of sudden changes in climate will likely to alter the species of composition of the plant and net primary production (NPP). According to the integrated observation of more than three decades and monitoring with a manipulative experiment of the temperature and the precipitation, in order to explore the impacts of the alterations in the climate on the plant community structure as well as functions of the ecosystem.

First of all, long-term climate warming from the year of 1983 to the year of 2014, which took place without systematic alterations in the precipitation, led to higher grass abundance and lower sedge abundance, but did bit impact aboveground net primary production.

Second is an experimental warming experiment which is conducted over the time period of four years, had no impacts on any aspect of the net primary production. On the other hand, drought manipulation (which is reduced precipitation by almost half, that is 50%), shifted net primary production allocation belowground without impacting the total net primary production.

Figure 3.6. Shifting plant species composition in response to climate change stabilizes grassland primary production.

Source: Image by Flickr.

Third, both drought treatments as well as experimental warming, aided by a meta-analysis at nine sites all across the plateau, increment in the grass abundance at the expense of the biomass of the sedges and forbs. This alteration in the functional group composition led to a deeper root system, which may have enabled the plant communities in order to accomplish more water. In this way, stabilize ecosystem primary production even with an altering climate.

Overall, according to the studies that demonstrates, changes in the species of the composition of the plant in response to the sudden changes in climate may have stabilized primary production in this high-elevation ecosystem, but it also produces a shift from aboveground to the belowground productivity.

Sudden changes in the climate will affect the structure and the function of the terrestrial ecosystem, more specifically in cold regions. The amount of the net primary production (NPP) in an ecosystem impacts the carbon feedback between the terrestrial biosphere and the atmosphere.

More and more experimental studies have explored that how plant community composition and the primary productivity will respond to the sudden changes in the climate. Nevertheless, several numbers of studies have done so. With respect to the long-term observations, in order to discover how alterations in the community composition that are triggered with the help of sudden changes in the climate can impact the functions of the ecosystem.

Alpine ecosystems are well thought out from the most sensitive ecosystems, with respect to the climate alterations. Near about 60% of the Tibetan Plateau is Alpine. Over the time period of around five decades, annual temperature on the Tibetan Plateau has elevated very drastically with a rate of two times as compared to the global average.

In addition to this, data shows that, from the year of 2000, the rate of warming can be increased even more as compared to the time span of the last four decades. Alpine grasslands are very vulnerable to the warming, also they are diverse, and harbor large stocks of the carbon in the soil that, if they released, they can take into the direction towards a positive feedback for climate warming.

The ubiquitous alpine grasslands on the Tibetan Plateau are very significant with respect to the livelihood of the native nomads and the potentially vulnerable to the sudden changes in the climate. Several numbers of the approaches have been taken, in order to cope up with the effects of the climate change on this high altitude that is alpine ecosystems.

Both process-based ecosystem models and the long-term time-series datasets of the Normalized Difference Vegetation Index (NDVI) recommends that increment in the vegetations over the time period of past several numbers of decades, nevertheless, with the pronounced geographical heterogeneity.

Even though, these kinds of approaches, while informative at large spatial scales, do not offer detailed fine-scale information or insights into the potential mechanisms. Long-term observations at the local plot scale have shown that net primary production in the alpine grasslands appeared to be very insensitive to the modifications in either precipitation or in temperature of the air from the year of 1980 to the year of 2000.

An observational methodology like the one that is taken in the study cannot explicitly address the underlying mechanism that explains why the net primary production was unresponsive with respect to the sudden changes in the climate change.

In the end, a number of manipulatives explaining experiments on the Tibetan Plateau, with the help of either open-top chambers or infrared heaters, have reported inconsistent outcomes, some showing the increment in the net primary production and the grass biomass. On the other hand, others are finding the decrement in the net primary production.

In addition to this, nevertheless experiments can be very useful in order to explain the mechanisms and providing the boundaries of the possible change. They cannot hold the legacy by the background warming and significantly, do not reflect reality.

Given these uncertainties, one vital way forward is to integrate the long-term observational studies with the experimental manipulations in order to examine the long-term patterns as well as the detailed mechanisms of the community and the ecosystem responses with respect to the climatic alterations.

In the present interval of time, according to several numbers of studies, and integration of around 32 years of observational monitoring of the alpine grassland ecosystems on the Tibetan Plateau with a manipulative experiment in order to discover the impacts of the climate change in the context of productivity of the grassland productivity and community composition.

Particularly speaking, researchers have evaluated that, whether it is net primary production and the plant community composition were sensitive to the sudden changes in the climate in this high-elevation, cold systems, and whether the changes in the plant community composition amplified or stabilized changes in the productivity. In the end, it is also explored that the possible underlying mechanisms of the impacts of the global warming on the primary productivity.

3.7. CLIMATE CHANGE EFFECTS ON FORESTS AND GRASSLANDS

The average annual temperature in the several regions of the United States has elevated over the time period of last ten decades. Most of this warming has occurred from the time of 1970 s. As a matter of fact, each and every decade from the time period of 1980 s, has set a new record for the warmest decade on the record.

The average temperature in the United States has been increased to more than two degrees Fahrenheit in the time period of last five decades. Increment in the temperature are having extensive impacts on the forests as well as the grasslands. These impacts are consisting of precipitation patterns, insects, streamflow, snowpack, and invasive plant species, and influencing drought, wildfire, and heat waves.

These are just some of the examples of the observed impacts of the sudden changes in the climate all across the United States, and most of the impacts are expected to lasts. And also, these impacts will be going to be elevated through the course of the century.

Emission of the greenhouse have been increasing more and more, and also, greenhouse gases are impacting the present conditions of the climate. The choices of the humans make about land use, energy, and the development will impact the future amount of the greenhouse gas emissions.

Figure 3.7. Climate change effects on forests and grasslands.

Source: Image by Geograph.

In order to anticipate the future impacts of the sudden changes in the climate, several kinds of instances with respect to the emission of the greenhouse gases in the future, also known as "pathways," that are used as inputs to the models of the climate. Each and every climate model, as an outcome, helps in stimulating the climate system in a different manner.

The future scenarios of the emission of the greenhouse gases are integrated with the several kinds of climate models result in a range of temperature and the precipitation outcomes from a warmer to much hotter, as well as drier to wetter. Generally, instances of higher emission of the greenhouse gases lead to the anticipation of more severe changes. On the other hand, cases with lower emission or pathways lead to the projection of the less severe changes.

The range of the emission scenarios and the outputs from the climate models represents most of the uncertainty in the future climate projections. The climate is certainly altering and will endure to alter or shift. But because researchers still have some unknowns about the future. Humans have to work with a plausible range of the future climates that integrates the variety of the emission pathways as well as the model projections.

3.7.1. Benefits/Stresses

Under the right set of situations or settings, some of the ecosystems may benefit from the alterations in the climate. These benefits can only last for few decades, or it could be possible that it will lasts for much longer. Nevertheless, it is more likely that the sudden changes in the climate will worsen the present stressors in an ecosystem or add new stressors.

Every so often, these kinds of benefits and the stressors that have been identified are a part of personal values and beliefs about how ecosystems should look and function. It is very important to think about where the advantages and the stresses can take place. In this way, humans will get prepare in order to take advantage of the sudden changes in the climate and overcome challenges from the climate change stressors.

3.7.2. Effects

There are several numbers of ways in order to describe the impacts of climate change on the forests and grasslands. In this topic of the chapter, the prime impacts on forests and the grasslands are integrated into four extensive categories. These categories are, wildfire, water, disturbance, and vegetations.

Each and every category may consist of both advantages as well as stresses to the ecosystem health and the function. The impacts may not take place in the isolation, and are, in fact, are very likely to interact with each other.

3.7.3. Water/Precipitation

Several numbers of various regions all across the United States have already experienced alterations and modifications from the historical precipitations: some of the regions are receiving more precipitation and more heavy precipitation, and others are becoming drier. Precipitation in several numbers of places has become more variable, every so often confusing the clear trends. Alterations in the precipitation patterns and the amounts impacts soil moisture, risk of wildfire, and level of groundwater.

Projections of the future precipitation show high variability, and every so often, several kinds of emissions scenarios depict opposite patterns in the rainfall through the course of the same season. Regions that are projected in order to experience more annual precipitation in the future may not necessarily benefit, because more of that precipitation is anticipates to fall through the course of intense storms.

This is already taking place in several numbers of regions, as an outcome in more frequent flooding and erosion. Some forests are experiencing longer dry periods between the heavy precipitation events. These dry periods can result in water stress in forests, and the heavy rains may not help in decreasing the water stress because more precipitation flows directly into the streams, in spite of filtering into the soil.

3.7.4. Water/Snow Cover

In the time period of last several numbers of years, snow cover across much of the United States has reduced in depth, extends over a smaller area, and melts sooner in the spring. Alterations in the snow cover and snowmelt have extensive impacts. Less snowpack integrated with the earlier melting that offers less insulation for plants and soil, exposing them to the frosts as well as freezing temperatures.

Early snowmelts also shift the timing of the runoff into the streams, with the large amount of flows taking place earlier, that is followed by diminished flows late in the growing season. Low flows in the summer and fall can take into the direction that can lead to the water stress for the plants, that can take into the direction towards the elevated risk of insect infestations and the wildfire, and impact the soil processes. Lower flows late in summer also result in the lower water volume and higher temperatures in streams, which may impact the population of fish in a negative manner.

3.7.5. Water/Flow

Altogether, the alterations in the precipitation patterns and snow cover impact the amount and the timing of the runoff as well as streamflow. Particularly speaking, rain-on-snow events, when rain falls on snow, are projected to become more frequent and intense as the temperature rises.

Warmer temperatures alterations through the course of winter precipitation from snow to rain, and the warmer rain filters into the snowpack, which produces melting, increment in the runoff and the danger of flooding as well as conditions of erosion.

Decrement in the summer seasonal flows and increment in the flash floods and short periods of high flows from more intense storms are anticipated impacts of the sudden changes in the climate through the course of the century.

3.7.6. Water/Sea-level Rise

Oceans absorbs more than 90% of the extra heat in the atmosphere from the emission that is produced by the humans. As the water heats up, it expands, contributing to the increment in sea level. Global sea levels have been elevated to near about 8 inches from the time period of 1880 s, and from the year of 1992, the sea level all across the globe has been increased two times as fast as the long-term trend.

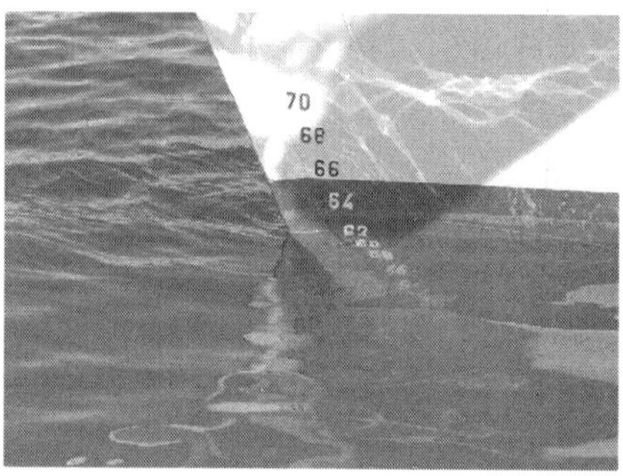

Figure 3.8. Increment in water/sea level rise.

Source: Image by Pixabay.

By the end of the century, sea level all across the globe is anticipated to increase by one to four feet. This increment, along with the projections for more coastal storms, will elevate the danger of erosion, flooding events, storm surge, and impacting the infrastructure as well as the coastal ecosystem.

3.7.7. Vegetation/Growing Season

The frost-free season, every so often is used to define the length of the growing season, has lengthened around the country since the time period of the 1980s, with the largest increase observed in the west. The first freeze is happening later in the fall, and the first thaw is happening earlier in the spring. The increment in the length of the frost-free and growing seasons is predicted to last in this century.

In the earlier time period, snowmelt and springtime have led to earlier emergence of leaves, or leaf out, and earlier flowering in several numbers of plants. Longer warm periods mean that the plants have a longer time to prosper, which can elevate the productivity of the forest and grasslands if the plants have a sufficient amount of water as well as nutrients. This could be a very beneficial impact of the sudden changes in the climate.Nevertheless, there may also be some negative aspects to a larger growing season.

3.8. CONCLUSION

In the conclusion of the chapter, various impacts of sudden changes in climate on the grassland production and soil carbon have been discussed. This chapter revolves around the discussion on effects of climate change on forests and grasslands.

It also discussed about the effects of increment in atmospheric carbon concentration and CO climate interaction. In this chapter, various effects on grassland biodiversity along with productivity due to sudden changes in climate have also been discussed. Towards the end of the chapter, the concept of shifting plant species composition in response to climate change stabilizes grassland primary production has been discussed as well.

REFERENCES

1. Climate change connection, (2015). *Grasslands Impacts | Climate Change Connection.* [online] Climatechangeconnection.org. Available at: https://climatechangeconnection.org/impacts/ecosystems-impacts/grasslands-impacts/ (accessed on 7 July 2021).
2. Conservation in a changing climate, (2020). *Climate Change Impacts to Grasslands.* [online] Conservation in a changing climate. Available at: https://climatechange.lta.org/impacts-to-grasslands/ (accessed on 7 July 2021).
3. Liu, H., Mi, Z., Lin, L., Wang, Y., Zhang, Z., Zhang, F., Wang, H., et al. (2018). Shifting plant species composition in response to climate change stabilizes grassland primary production. *Proceedings of the National Academy of Sciences,* [online] *115*(16), 4051–4056. Available at: https://www.pnas.org/content/pnas/115/16/4051.full.pdf (accessed on 7 July 2021).
4. Parton, W., Ojima, D., & Schimel, D., (1994). Environmental change in grasslands: Assessment using models. *Climatic Change,* [online] *28*(1, 2), 111–141. Available at: https://link.springer.com/article/10.1007/BF01094103 (accessed on 7 July 2021).
5. Parton, W., Scurlock, J., Ojima, D., Schimel, D., & Hall, D., (1995). Impact of climate change on grassland production and soil carbon worldwide. *Global Change Biology*, [online] *1*(1), 13–22. Available at: https://www.researchgate.net/publication/233714480_Impact_of_climate_change_on_grassland_production_and_soil_carbon_worldwide (accessed on 7 July 2021).
6. Science Net Links, (2020). *Grasslands and Climate Change.* Science NetLinks. [online] Sciencenetlinks.com. Available at: http://sciencenetlinks.com/lessons/grasslands-and-climate-change/ (accessed on 7 July 2021).
7. *SCRIPT Climate Change Effects on Forests and Grasslands: What You Need to Know*, (n.d). [e-Book] Available at: https://www.fs.usda.gov/ccrc/sites/default/files/CC_Effects_Narration_Script1.pdf (accessed on 7 July 2021).
8. Van, O. M., Bellocchi, G., & Höglind, M., (2018). Effects of climate change on grassland biodiversity and productivity: The need for a diversity of models. *Agronomy*, [online] *8*(2), 14. Available at: https://www.mdpi.com/2073-4395/8/2/14/pdf (accessed on 7 July 2021).

9. Ze, H., Xiangzheng, D., Xinliang, X., & Wei, S., (2020). *Grassland Ecosystem Responses to Climate Change and Human Activities within the Three-river Headwaters Region of China*. [online] Springer Nature. Available at: https://www.nature.com/articles/s41598-018-27150-5 (accessed on 7 July 2021).

Chapter 4

Environment and Crop Production

CONTENTS

In this chapter, the environment and crop production has been discussed in detail. The heavy metal stress and crop productivity has also been done in detail. The effect of industrial pollution on crop productivity has also been done.

In this chapter, there is also a detailed discussion about pesticide tolerance and crops production. Finally, climate change and plants has been discussed in detail, and factors like temperature alterations, changes in precipitation regimes have been discussed in detail.

4.1. INTRODUCTION

The field of agriculture have progressively focused on the creation of new technologies to meet the increased demand for food and supply over the years due to the drive to increase production policies. Its benefit for farmers is twofold; it has allowed for significant yield increase while lowering costs for farming.

More productive farming methods have emerged over time. Some recent developments such as the use of fertilizer and pesticides have increased the capability for yields with less labor and machinery, and farming practice have helped to lower the cost of land use, as well as development of hybrid varieties and progress in machine technology have provided for improved yield.

The positive impact of intensification of agriculture on the development has, however, has not been completely offset by the damage it has caused to the environment. The intensification of agriculture may have a negative effect on all three of these, namely, water pollution, loss of habitat, and desertification of land features, the environment, and deterioration of the landscape.

Recognizing a rise in the critical resource base level while maintaining productivity as a high a substantial natural base is needed, the agriculture industry has expanded its attention to environmental performance, improving its effectiveness and reducing its damaging environmental impact. Because the agricultural production depends on both natural resources and the environment, it is more complex than one thought.

Agriculture is a big user of land and water resources but has to protect and enhance the resources in order to continue to be economically feasible. Agriculture may be a significant source of waste and pollution, but it is also a significant for its resource conservation and recycling abilities. There are a lot of external influences on the environment that go hand in hand with agriculture, but few of them happen to the farm itself. It is often an issue that impacts individuals and regions rather than national and global policy, although some issues have global consequences. During the next half-century, demand for food by a wealthier and larger global population will put great pressure on agriculture.

The methods and techniques that use living organisms to create or modify products employ biological materials are biotechnologies. The biotechnology that uses conventional techniques such as plant and animal breeding is well accepted to expand the current range of viii qualities and yields is capable of doing so. Scientists now alter genetic material in living organisms using modern biotechnology that uses the terms to achieve greater evolutionary changes.

The enzymes derived from genetically modified microorganisms are commonly employed in industrial settings, but the long-term effects of using microorganisms and GMOs are unknown. In addition to this, agriculture has caused a rise in greenhouse gas emissions, and a loss of forest acreage due to conversion to farmers planting crops. The reality is that climate change could impact and permanently damage many natural resources that are vital to the agricultural economy.

It is possible that moderate global warming will slightly increase crop yields, but on the whole negative impacts will grow. Agricultural disasters, including floods and droughts, become more severe, which could severely impact rural communities. It is also possible that climate change will serve to expand the geographical range of diseases, and that in turn, support the spread of harmful organisms like insects and weeds.

The higher levels of ultraviolet and gamma-rays in the atmosphere with nitrous oxide emissions causing the soil to use up either during their production or use also causes increased levels of nitrogen fertilizer, so it results in more frequent ozone destruction. There is an entire chapter that deals with the impact of gamma rays on production, and it is an extremely interesting subject that is completely new to us. Many of the world's diseases continue to go undetected because the food supply has been increasing while the remaining 85% of the world's disease burden has been going untreated.

Many groups in the population are found to be deficient in protein, micronutrients, and vitamins. Once the political frameworks are established, agricultural research and policies must be developed and procedures for improving dietary diversity, as well as promoting food processing will be instituted. The different health concerns that apply to agricultural products include the presence of pesticides and toxic or contaminated metals inks.

Taken as a step further fact, agricultural development was made up of many various projects which promoted productivity and was focused on harvesting natural resources, but failed to reflect the overall effect on agriculture, the relationships between agricultural activities and the local ecosystems. The challenges of producing healthy crops and quality food grains, for example, must be solved if soils are polluted by heavy metals and industrial effluent, but the creative efforts of all concerned are required for their realization. The wheat production has staggered in South Asia in recent years, due to the adversity of biotic and abiotic stresses in particular, because new varieties are unable to significantly increase their yield capacity.

The wheat breeders are unable to develop unique genotypes of great enough in this day and age and have to focus on better ways to use genetic diversity that exists in older varieties to increase it. The fact that many traits seen in plants which are found in wheat's wild relatives as well as those related species do provide an extraordinary reservoir of stress and biomass, and yield, and also exist as the plant and photosynthesis resilience for new varieties of higher yields to be developed to handle the global warming, an optimum environment is necessary, but this is only an obstacle for the expansion of older varieties of farming that are less sensitive to environmental change.

The wheat production should be considered in the context of changes to its overall productivity, which include the introduction of new genetic alien strains, to avoid new problems resulting from excessive stress. Information concerning the role of seed quality also describes the advantages that can also focus on a farmer's practice of following higher yield.

4.2. CROP PRODUCTION AND ENVIRONMENTAL PROTECTION

Crop development has always had the tendency to have an effect on the ecosystem because croplands are inextricably connected to the soil and the environment. Both agricultural activities have always had an impact on soil quality, but the degree of that impact depends on the context. Although new farming practices are employed and farm managers are being concerned with keeping the soil well fed, a positive impact on the climate may be achieved.

Historically, farmers have realized that they could add nutrients back into the soil after they have harvested their crops but were too preoccupied with making a profit to recognized the value of doing so. Lithic soils eventually exhausted, this land, which resulted in reduced crop yields and accelerated deforestation, polluted rivers, and the streams, and caused the area to have a growth in turbidity. When the soils wore out, farmers and their families pushed on to other areas and started the arduous task of doing the same.

Too many farmers around the world are finding themselves in situations where they want to provide enough food to eat in depleted soil and water-deficient soil management practices, further driving down their soil fertility. Because of their tendency to be cultivated, especially because of their slope, soils tend to lose a lot of their topsoil, which causes both surface and groundwater pollution the nutrients are removed from the soil, and subsequently replenished by environmental factors. As the soil is eroded, the ecosystem gets damaged; as it eventually recovers, the world gets damaged.

On the other hand, current agricultural activities have shifted the nature of the ecological balance, but not eliminated it completely. Today's farmers are better guardians of soil and water than their predecessors, having learnt the lessons of past abuses, and future problems, and appreciate their position in conservation. Many modern farmers consider that agricultural management and fertilizer use must bear in mind a long-term view, not just this year's goals, in order to do the best they can for a long-term future advantage.

The people are going hungry and the world over, making an effort to provide more food, putting more stress on the food-producing capacity of farmers to do to meet their demands. To guarantee the food supply of an extra one and a million people five days' daily meals is just as difficult as it is additionally protecting the atmosphere from deterioration.

Farmers must also use fertilizer and manure to increase the nutrients in the soil in order to achieve high-quality yields. With the rapid increase in species diversity, the biodiversity on Earth has become precarious, and susceptible to losing its equilibrium, though concurrently they are devoting greater effort to maintaining it.

They will get a better return on our investment as crop yields rise, but the amount of fertilizer that we use will as well. This has been accounted for by the fact that it takes more units of crops to apply the same amount of fertilizer that is being extracted per unit of soil. When this happens, many of the nutrients that are added to crops are captured by the soil rather than the water, allowing them to remain in the soil or ending up in streams.

This ensures that more ecologically sensitive land will be put into agricultural production, protected from development, used for biodiversity, and diverted to sustainable water habitat and other uses that better support the ecosystem.

Often, it is a matter about who has to pay for water and air contamination or collateral harm to land and infrastructure that is decided, which means there is no simple divide between protecting the atmosphere and causing trouble. As a result, modern farmers are conducting experiments in food production to help high yielding crops thrive while safeguarding the ecosystem is becoming more and attempting to preserve it.

The people and their families and breathe the air, consume the food, and consume the water that is produced on the farm to everyone frustration, there is no way to reconcile the need for agricultural production with conservation.

Figure 4.1. Crops Production get effected due to air pollution.

Sources: Image by Pixabay.

4.3. HEAVY METAL STRESS AND CROP PRODUCTIVITY

A potentially harmful substance can be presented as a trace metal or a toxic heavy metal. The trace elements in soils can be classified as those found in quantities described as 1 mol m^{-3}. As a loosely described as metallic elements, metal compounds have atomic weight higher than 5 but lower than 20, represents a wide class of metals and specific gravity greater than 5 (or the metals that are out of the metal family of the alkali metals (Rascio and Navari-Izzo,2011). The few heavy metals are the ones that have an atomic density of at least 6 g/cm −3, except for As and Se (Park et al., 2011).

Many transition metals, metalloids, and actinides can be dangerous to living things at low doses. This has however been questioned by the International Union of Pure and Applied Chemistry (IUPAC) in their article on the subject of misinterpreting the terms because of the opposing meanings and absence of a coherent theoretical rationale for classification for heavy metal.

In this context, the word heavy metal refers to certain plants that have been tested and found to be poisonous or toxic at extremely high concentrations. Of the naturally occurring elements, about 53 are deemed to be heavy metals, but these metals do not have any critical functions in plants. Heavy metals have certain distinctions when it comes to the role they play in plants and the role they play in the biological importance of heavy metals vary (Kavamura and Esposito,2010). The researchers believe that the role of a plant is positively influences its production.

Some of these heavy metals, such as zinc, copper, manganese, selenium, and chromium, are necessary for normal, and thus produce biologically correct effects on plants (Kavamura and Esposito 2010; Rascio and Navari-Izzo 2011; Salla et al., 2011).

Furthermore, however supraspinal concentrations of these elements and those with no metabolic activity, like the Zr, Sb, As, Pb, Hg, and Cd limit crop productivity (Park et al., 2011; Rascio and Navari-Izzo 2011; Shahid et al., 2013; Foucault et al., 2013; Xiong et al., 2014; Pierart et al., 2015).

The Agency for Toxic Substances and Disease Registry (ATSDR) has classified these last four additional radioactive metals (As, Pb, Cd, and Hg) as being the most toxic due to their toxicity, the extent of incidence, and exposure potential to humans.

Thus, soil pollution of heavy metals can result in increased metal concentrations in plant tissue, which decrease plant yields and increase risk to the health of exposure to these two toxins in both animals and people (Järup 2003; Foucault et al., 2013; Sabir et al., 2015; Saifullah et al., 2015).

It is a number of researchers have documented the appearance of heavy metals in food plants, especially in soil that has been polluted (Uzu et al., 2009, 2010, 2011a; Nabulo et al., 2010, 2011). In numerous parts of the world, metals, including Cd, Pb and Zn were found to be present, making the soil not suitable for farming (Singh et al., 2010; Nabulo et al., 2010; Uzu et al., 2010, 2011a; Bai et al., 2011; Foucault et al., 2013; Shahid et al., 2014a).

Several studies indicated that crops grown near a town's or city's population limits can show detectable levels of pesticide residues (Yang et al., 2011; Uzu et al., 2011b; Xiong et al., 2014).

In regard to the assessment of Lead concentrations in Nigerian vegetables, finding that had results above the normal in three cities in total (Adekun et al., 2009). In the same manner, Singh et al. (2010) has found a threat of Cd and Zn poisoning due to food grown in India, caused by ingesting contaminated vegetables that indicate soil contamination of metals.

The researchers have expressed concern about the significant hazards caused by frequent exposure in Iran to cadmium, zinc, and lead polluted sites because of power plants and industries involved in metal production (Seilsepour and Bigdeli 2010).

A study published in China (Zang et al., 2009), as well as two from Australia (Kachenko and Singh, 2006) and one from France (Schreck et al., 2011, 2013; a similar finding is also recorded in Foucault 2013), found heavy metal beyond the food limits in vegetables grown near metal smelting plants.

Plants become more metal-toxicated as a result of heavy metal accumulation in the soil and thus decrease crop productivity, as well as impeding uptake of water and minerals and nitrogen, or increased metal uptake can result in excessive metal concentrations that restrict the growth of plants by interfering with photosynthesis and other soil processes (Cambrolle et al., 2011; Dirilgen, 2011; Shahid et al., 2014; Buendía-González et al., 2014; Austruy et al., 2014; Hasanuzzaman et al., 2014).

When heavy metals are exposed to soil, it may be less able to support crops and will, and, in serious situations, lead to plant death. It may disrupt the structures of grana, minimize root and sparkling growth and decrease chlorophyll biosynthesis, affect breathing and photosynthesis (Cenkci et al., 2010; Bah et al., 2010; Pourrut et al., 2011, 2013; Ali et al., 2011; Austruy et al., 2014). In many plants, extensive cell death (necrosis) and programmed cell death (apoptimum) also coexist due to metal toxicity.

The generation of reactive oxygen species (ROS), disruption of cell redox status, is considered to be a significant contributing to heavy metals toxicity in plants (Duquesnoy et al., 2010; Deng et al., 2010; Antolín et al., 2010; Gill and Tuteja 2010; Körpe and Aras 2011; Pourrut et al., 2011; Shahid et al., 2014c; Ibrahim et al., 2015). The harm caused by these metals to normal biological processes may be either proactive or due to their proximity to critical ions (Kramer et al., 2007). The frequent use of heavy metals has led to the evolution of various protective mechanisms that help plants deal with the stresses. These pathways support cell redox crops by balancing damage caused by oxidative stress and maintaining crop productivity (Xu et al., 2011; Xiong et al., 2010; Buonocore et al., 2010; Shahid et al., 2014c; Sabir et al., 2014; Sabir et al., 2015).

4.3.1. Heavy Metal Effects on Crop Productivity

Although within the plants, heavy metals are capable of causing changes in morphology, physiology, and biochemistry, making them less productive, expanding would most likely result in their being neutralized and facilitating the return to normal. But metals promote toxicity, regardless of the age of the stage of the plant or the level of stress under which the target organism, the organisms, exposure to heavy metals, the severity of the treatment, and application of metals, and wherein the plant the treatment is performed.

4.3.1.1. Effects of Heavy Metal on Germination

As in the life cycle, physical and biotic (or micro or macro environments) associations like germination are generally the most influential forces on the metabolic growth of plants, but these are often altered by a myriad of organic and metabolic (or biotic or environmental) compounds (Moosavi et al., 2012).

The type of metal that has an impact on seed germination is mostly dependent on how far it gets into the seed and how far it can penetrate (Ko et al., 2012; Márquez-García et al., 2013). The same amount of metal added has differing results in various plants based on different structures in seed coat anatomy (Munzuroglu and Geckil, 2002).

The metallic elements such as copper and cadmium water absorption precede the development of a plant's ability to germinate (Kranner and Colville, 2011). The toxicity of the soil can be tested by seeing how the germination rate has increased (Munzuroglu and Geckil, 2002).

Most seeds store reserves for respiration are blocked after penetration into the seed coat, but under normal conditions, some seed production is capable of breaking down and providing metabolites in order to replenish those that is halted by metal toxicity (Ko et al., 2012).

The effects of various levels of Cu, Mn, Ni, and Zn levels on germination and seedling sizes in *Atriplex halimus* and *Salicornia ramosissima* have been studied by Marquez-García et al. (2013). The germination rate was decreased in *A. halimus* by all the metals and *S. ramosissima* was reduced by Ni only. The increased Cr concentration decreased *L. perenne* germination rate (Chigbo and Batty, 2013). This decrease was found only at concentrations of more than 9.4 ppm Cr.

The reduced germination was likely due to the presence of insufficient food reserves and the fact that it was caused by oxygen uptake problems. These metals may be taken up by roots and through the soil or by rain or dust (Jozic et al., 2009).

The study found that the plant roots in road verges (as opposed to control sites) had up to four times the amount of iron, copper, and molybdenum, and one time the amount of manganese in them.

Even though the seeds are fine, they are particularly vulnerable to metallic traffic pollutants. We found decreased seed viability and a rise in the number of irregular embryos in plants developing along highly traveled highways (Stvolinskaya 2000).

Ganatsas et al. (2011) found, even after pretreatment, that the *Pinus brutia* seed collected near the poisoned road did not germinate. The seeds gathered over a prolonged period of time on the other side exhibited natural behavior.

4.3.2. Effects of Metals on Accumulation and Remobilization of Seed Reserves during Germination

In the last decades great amount of understanding is now known about seeds, one of the areas that has been the most thoroughly studied recently in plant physiology (Bewley,1997). The researchers have studied the distribution, accumulation, and dispersal of heavy metals in a variety of plant species (He 2010; Lange-Hesse et al., 1994; Mengchang et al., 2000; Wang et al., 2014).

The success or failure of developing seedlings during early ontogeny is dependent on the restoration of transcriptional and translational machinery (Finkelstein 2010; Mei and Song 2008) and a number of other factors, including the successful remobilization of stored seed reserves (Nadeem et al., 2011, 2012, 2013) that accumulate in various seed compartments during the ripening period (Malan and Song 2008); Seedling root growth (Enns et al., 2006; Zhu and Lynch 2004); soil, water, and temperature factors (Louarn et al., 2008; Manz et al., 2005; Miller 2001); and metal concentrations (Chigbo and Batty 2013; Ko et al., 2012; Kranner and Colville 2011; Márquez-Garca et al., 2013; Wang et al., 2014).

The seed germination stage is influenced by metals, specifically Cd, Cu, Cr, and Mn, specifically in the atmosphere (Chigbo & Batty, 2013; Kalai et al., 2014; Léon et al., 2005; Ko et al., 2012; Márquez-García et al., 2013).

In comparison to ordinary Ni concentrations (Léon et al., 2005), higher Ni levels were induced by the remobilization of macronutrients, while the remobilization of stored barley seed phosphorus reserves was recorded as a result of a decline in phosphatase, alkaline phosphatic, and phosphatase activities in endosperm, mainly due to Cd and Cu (Kalai et al., 2014) concentrations. Moreover, Cu and Cd also increased the level of soluble protein and sugar and induced the aggregation of proline, mostly in radicles, also at the lowest dosage (Kalai et al., 2014).

The reduction in the germination of barley seed after exposure to Cu or Cd is not attributable to a reduction in seed tissue water absorption but may be due to a failure in endosperm reserves (Kalai et al., 2014). Increased concentrations of Cu, Cr, Co, Ni, Mn, Pb, and Zn metals decreased germination percentages in a variety of crops (Chigbo and Batty 2013; Márquez-Garca et al., 2013; Wang et al., 2014), and this decrease in germination rate was attributed to limited oxygen intake and physiological disturbances in the supply of food supplies (Chigbo and Batty 2013; Márquez-García et al., 2013).

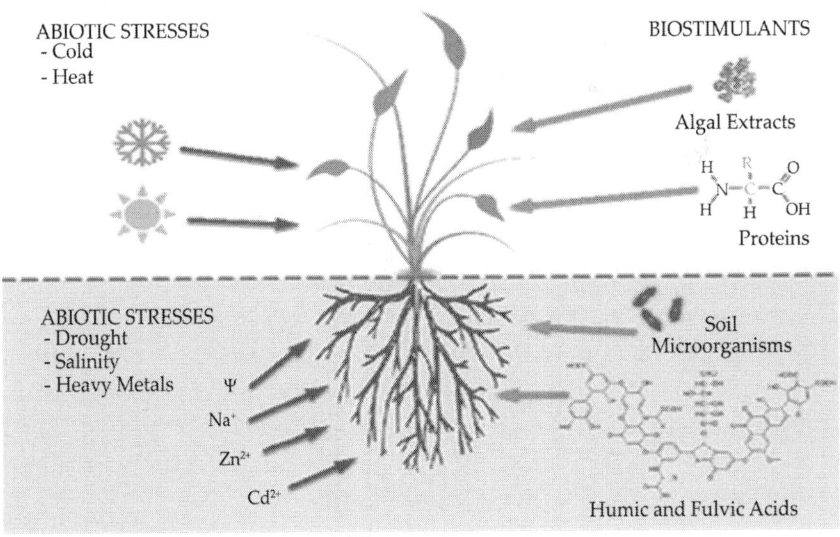

Figure 4.2. Heavy metal effect on crops.

Sources: Image by Wikimedia.

4.4. EFFECT OF INDUSTRIAL POLLUTION ON CROP PRODUCTIVITY

The agricultural crops may be damaged by contact with toxic or toxic gases in addition to other air contaminants. As a result, these studies have shown, injury manifests itself in both color change and crop yield changes, from the plant death can happen anywhere from foliage blemishes to stunted growth to early demise. The nature and the level of the pollutant itself does not determine how the illness develops and how it appears, but rather on various other influencing factors.

It also takes into consideration things like the degree of exposure to the pollutant and the type of the species and the plant and its level of growth and also categorizes them based on certain preconditions, such as whether they are facilitative or adverse to pollutant to plant damage.

4.4.1. Air Pollution Problems

Generally, the damaging effects of vegetation in cities are from localized air contaminants, including atmospheric pollutants that come from the surroundings. Examples of local contaminants, those which originate from a single point of production and end-use, cause a distinct and confined region of injury and pollution. The contaminants include Sulphur dioxide, fluorides, ammonia, and airborne particles are found among the common pollutants.

The most common contaminants consist mostly of oxidants. A major component of oxidants ozone is released in the atmosphere through a series of reactions that include the combined oxidation of nitrogen oxides and reactive hydrocarbons present in automotive exhausts and fossil fuels as well as well as the partial oxidation of hydrocarbons in vehicle gas engine and vehicle emissions.

So, this mechanism can only occur in the presence of sunlight, which is known as photochemical reaction. Not only in cities but in rural areas as well, oxidation in the air can cause plant damage and can spread over hundreds of square kilometers.

4.4.2. Effects of Air Pollution on Plants

There are various ways in which air pollution can negatively affects plants. The Necrotic tissue injury can be present for a brief period, and the leaves may exhibit chlorosis which yellowing symptoms or may occur gradually and affect the leaves and remain undetected. There may be a decrease in different components of the plant development of certain parts. The treatments generally destroy the plant species directly, but typically do not kill them immediately because of the damage that occurs over and over again.

4.4.3. Oxidants

In the oxidant smog complex, ozone is the primary pollutant. One of the first detections of its impact on plants was in Los Angeles in 1944. The damage done to vegetation in Canada's south-western and central regions has since been recorded as ozone damage and can be documented in the other parts of the country as well. When it is warm, the levels of ozone vary greatly, in particular during the growing season in July and August.

People in the lower Great Lakes regions can often find higher levels of ozone levels during the months of fall when prevailing winds from the south are a carry high amount of it through urban and industrialized areas of the US. Of particular concern is the high background ozone levels, which is generated domestically and by localized sources. Any other plants that are also vulnerable to decline are cucumber, grape, green bean, cabbage, leek, rutabagas, tomato, squash, lettuce, spinach, sweet corn, and corn. Endive, pear, and apricot are resistant varieties.

The upper surface of the infected leaves is characterized by ozone symptoms which seem to spot, bromize, or bleach the leaf tissues. Although generally, yield losses are associated with leaf tension, however, they may also occur in the absence of any apparent indications of pollution tension. Additionally, crops can maintain or even improve their leaf photosynthetic ability despite any apparent foliar damage.

Other environmental and plant factors, as well as health effects can determine susceptibility to ozone damage. Higher humidity in the plant's surroundings, greater abundance of nutrients, and water in the soil all enhance vulnerability. One of the criteria that play a significant role in determining injury growth on large leaves is the stage of maturity. The smallest leaves are resilient. With each increase in size, they are successively vulnerable at both middle and base. As soon as the leaves reach full maturity, they harden up.

4.4.4. Sulfur Dioxide

Major sources of sulfur dioxide are coal-burning operations, especially those providing electric power and space heating. The Sulfur dioxide emissions can also result from the burning of petroleum and the smelting of sulfur-containing ores. The Sulfur dioxide enters the leaves mainly through the stomata (microscopic openings), and the resultant injury is classified as either acute or chronic.

The acute injury is caused by absorption of high concentrations of sulfur dioxide in a relatively short time. The symptoms appear as 2-sided (bifacial) lesions that usually occur between the veins and occasionally along the margins of the leaves. The color of the necrotic area can vary from a light tan or near white to an orange-red or brown depending on the time of year, the plant species affected, and weather conditions. The recently expanded leaves usually are the most sensitive to acute sulfur dioxide injury, the very youngest and oldest being somewhat more resistant. Chronic injury is caused by long-term absorption of sulfur dioxide at sublethal concentrations. The symptoms appear as a yellowing or chlorosis of the leaf, and occasionally as a bronzing on the undersurface of the leaves.

Different plant species and varieties and even individuals of the same species may vary considerably in their sensitivity to sulfur dioxide. These variations occur because of the differences in geographical location, climate, stage of growth and maturation. The following crop plants are generally considered susceptible to sulfur dioxide: alfalfa, barley, buckwheat, clover, oats, pumpkin, radish, rhubarb, spinach, squash, Swiss chard, and tobacco. Resistant crop plants include asparagus, cabbage, celery, corn, onion, and potato.

4.4.5. Fluoride

The burning of coal, the manufacturing of brick, tile, frit enamel, glass, ceramics, and steel and production of aluminum and steel; and in production of hydrofluoric acid, phosphate, and fertilizer chemicals, result in release of fluorides in the atmosphere. The preferred destination of fluorides in broad leaves in grapes and on monocotyledonous leaves tips in gladiolus. The small or minimal damage happens at the intake or place of absorption, while more severe harm takes place at the absorbent (i.e., the edges or tips) levels.

At first, it appears as a grey, or greenish-colored water-rot, and becomes darker, until it is eventually reaching a darker reddish-brown. The Monocotyledonous leaves have a dramatic expansion of the necrotic regions with continued exposure. They can expand downwards and in an areolate expansion of necrosis spreads around and gets worse with long-term exposure.

The studies have shown that Apricot, Barley, Peach, Gladiolus Corn, Plum, and Tulip to be most sensitive to fluoride. The list of hardy plants includes alfalfa, asparagus, beans (snap), Brussels sprouts, cabbage, cucumber, carrot, pepper, beet, sweet potato, and turnip, as well as the sturdier members of the legume family (beet, pea, rice, soybean, lentil, or bean) and rye.

4.4.6. Ammonia

More commonly, recent occurrences of damage to plants from the use of anhydrous and aqua ammonia fertilizers have been found related to fertilizer spills in the areas of application, transportation or storage, or application. The areas around these incidents have been reported to suffer severely from these incidents, due to the ammonia released into the atmosphere for short periods of time and their severity.

Typically, a whole system expression occurs on affected the plant life stage, which may take several days to grow, and sometimes consists of whitened or rotting fleshy lesions. The grasses are also degraded in the reddish, intervinery or black upper soil.

As an exception, serious damage to fruit trees rarely has an effect on new leaf development, an uptick in their leaf production does not appear as a regular event occurs. Apple, barley, beans, clover, raspberry, and soya are among the most sensitive plants. These plants have high resistance include alfalfa, beet, tomato, chive, carrot, cucumber, corn, and eggplant.

4.4.7. Particulate Matter

The natural respiration and photosynthesis processes within the leaf may be induced by particular materials like cement particles, Magnesium-limestone, and carbon soot, accumulated on plants. The combining of thick crust and the alkaline toxicity caused in wet weather can cause chlorosis and death of the leaf tissue.

When using dusts added, such treatments can also impair the natural functioning of pesticides and other agricultural chemicals that are applied to vegetation. Additionally, aggregation of alkaline ions in the soil can also increases the pH, making it more alkaline, resulting in unfavorable for plant development.

Figure 4.3. Industrial fumes are a major source of air pollution.

Sources: Image by Pixabay.

4.5. PESTICIDE TOLERANCE AND CROP PRODUCTION

4.5.1. Pesticides and Pollinator Decline

Many insecticides have the further negative effect of endangering wild habitats, for example, bees, but can also affect the pollination of valuable fruit and vegetable crops like nut trees and berries. Following the development of more nontoxic alternatives to neonic alternatives, neonicotinoids (or neonics), a common class of pesticides, have increased in popularity due to their relative nontoxicity to mammals and humans.

However, an expanded set of research compares neonics to wild and domesticated pollinators; in controlled trials, bumblebee colonies exposed to neonics evolved steadily and produced fewer queen colonies than those without exposure.

And honey bees studies showed that neonics decrease comprehension and speech impairment. The results of the researchers are much more disconcerting because honey bees seem to choose food supplies that include neo-nicotinoids. Neonics do not only collapse the honey bee colonies, but when paired with other stressors, colonies such as varroa mites face a significant danger, such as the hundreds of miles to pollinate various crops in our urban food system and large, monocultured deserts.

4.5.2. Pesticide Resistance

The evolution of resistance in many insects, weeds, along with that of course, was aggravated by the excessive use of pesticides on farms. At the same time, all times, all living things have an inherent variability of their development. The farmers only use the same pesticides for a limited time, so the majority of the population that is vulnerable to the pesticide is eliminated.

However, natural genetic variation might imply that some pests might be immune to the pesticide, meaning that they are unaffected by it. Where a large amount of the same pesticides is continuously added to a population, these individuals rapidly accumulate pesticide resistance and soon result in entire populations becoming resistant.

When several herbicides are applied to an acre's worth of weed seed varieties develop a tolerance, these superweeds become a substantial problem for growers, who do not have the knowledge to deal with them.

The crop herbicide resistance is on the rise, and there could be several varieties of herbicide-resistant weeds in the fields, which will make it much more difficult to eradicate. The discovery of 25 additional species of weeds that have developed resistance to herbicides, already discovered by American scientists, is taking place all over the country.

Ten billion US dollars loss a year is projected on the American farmers are due to pesticide resistance issues that cost their crops. Instead of dealing with the causes of the issue, but without controlling or reducing resistance, agrochemical firms simply manufacture new formulations of numerous herbicides.

The chemicals may contain older, more often toxic, and carcinogenic products, such as atrazine and 2, 4-D; on occasion this type of reformulation is projected to increase the amount of chemicals applied to crops by as much as much as 700% by 2020.

Furthermore, the use of pesticides over time seems to cause resistance, meaning that farmers have to apply different formulations of pesticides constantly in order to stay on the treadmill of buying products. Excessive application of insecticides also leads to the development of resistant nontarget species, just as overuse on factory farms encourages antibiotic-resistant infections. Thus, this creates tremendous problems for public health experts who are working to help combat malaria and other such diseases.

4.5.3. Pesticide Drift

Depending on how they are administered, pesticides can inadvertently damage other agricultural products, residential or nonagricultural ones. An example of this is when a new herbicide developed for the field was given the OK from the EPA in 2017 and permitted for aerial spraying of crop seeds that were immune to a toxin was dicamba-resistant.

Many growers, who had assumed that Dicamba will be a useful against resistant glyphosate, felt pessimistic when the issue was addressed; this remedy proved to be volatile, even when administered according to the directions, it would return and destroy or damage every plant it is genetics wherever it is located. Because of drift problems, 3.6 million acres of nonacre soybeans and crops, as well as fruit trees, and vegetables in 25 states.

If a farmer does not want to cultivate pesticide-resistant crops and a pesticide drops over from a neighboring field, it will destroy or kill the agricultural crop and cause a bitter clash between farming groups. Instead of refusing to buy chemical and subjecting the producer to a regulatory fine, soybean farmers planted a new and more tolerant variety. Because of this, there was an increase in the amount of chemical being sold and use and the chances of it contaminating the fields, but a financial gain for the manufacturer.

Also, organic cultivation can be affected by pesticide fallout; residue can also lead to crop losses, thus contamination as well as and subsequent certification problems. Hence, farmers are losing out on the already higher cost of organic production, and other farms are restricted from paying equally for any additional expenses associated with it.

Figure 4.4. Honey bees are being wiped out due to the large scale use of pesticides.

Sources: Image by Pixabay.

4.6. CLIMATE CHANGE AND PLANTS

4.6.1. Increased Atmospheric Concentration of Carbon Dioxide

Although the increased accumulation of carbon dioxide in the environment affects certain plants differently, all plants are vulnerable to the effect. The photosynthesis is dependent on naturally occurring levels of carbon dioxide in the atmosphere. This should be good news for these crops because the increased levels of carbon dioxide in the environment should help fertilize them if all conditions remain optimal.

This phenomenon is unlikely to take place under favorable circumstances, due to ample water and warmth during the season. However, nitrogen shortage or extreme conditions of temperature during the season can have an opposite effect.

The nutritional value of leaves, root, stem, stem, and tuber plants are reduced with carbon dioxide as a lotion or however, are as nutrients, minerals and trace elements like zinc and iron, are far more specifically less prevalent (Taub et al., 2008; Loladze, 2014).

It has been shown that plant growth at higher CO_2 levels results in lower stomatal conductance and transpiration. In other words, plants use water and nutrients more efficiently, so that their biomass has a lower nutrient content. The increase in human and crop consumption has come with the rise in the insect population, which is now at an increased level where they consume as much as their own weight in crops in order to receive their daily ration of food (Hatfield et al., 2011). This can be done because the plants are capable of raising the carbon dioxide levels in their leaves until they begin to photosynthesize. Therefore, higher atmospheric levels of carbon dioxide do not benefit plants in natural conditions. However, under moisture stress conditions, most plants lose less humidity and less moisture in their yields (Simpson, 2017).

4.6.2. Temperature Alterations

Most of the many ways in which temperatures can adapt is to a rise or fall is by variation in average; the other is by variation in the night and day and the temperature versus the rest of the year. This is consistent with the general observation that crops are more vulnerable to heat during the reproductive growth period and when this one is taking place, which coincides with the stages of seed maturation (Hatfield et al., 2011).

Although it is tied to plant physiology and the reaction of plants to temperature changes is highly specialized and dependent on both on the phototropic behavior, which causes growth, and on developmental changes, which is involved in making new plant forms.

Some types of temperature and plant stress impact how long a crop production lasts, while others stress the total production. This change in temperature will influence the developmental parameters for each species, with a variation in each differing intensity. It will be essential to provide a different kind of reaction to these various shifts.

It is important though perhaps not sufficient to determine how a species will react to new temperatures as known limits are imposed, but it is useful but not necessary to have knowledge of how it has reacted in the past. Here are some of the phenological findings of the few crops that were examined over a long enough period of time to do important analysis:

- It is usually the seasonal rise in temperature that plants use more energy for respiration for maintenance and less for growth. With a 1°C increase in average temperatures, yields of the major food and cash crop species can decrease by 5–10% (Lobell and Field, 2007; Hatfield et al., 2009).

- When it increases with the seasonal climate change, the rate of an organism's energy use generally goes up as much as its need for metabolism goes down, so it uses more energy for respiration and less for growth. A 5–10% increase in yields of the most important staple food and cash crop production is anticipated to decrease by one°C with an above-average temperature rise (Lobell and Field, 2007; Hatfield et al., 2009).
- The whole process of plant growth is brought to a conclusion even faster due to the increased temperature (Hatfield et al., 2011 As opposed to ordinary breeding, which has to be done quickly, there is a greater possibility of lower yields and shorter life (Craufurd and Wheeler, 2009).
- This implies that plants are more resilient to higher temperatures than crops, or that photosynthesis in general seems to be greater with plants (Lipiec et al., 2013).
- Any temperate plants must be allowed to undergo a cooling cycle of at least two weeks of dormancy before they are able to bloom again. when a plant may not have enough chilling, it will only be able to grow vegetatively or only be able to produce a limited amount of fruit.
- Fruits and vegetables are also experienced changes in marketability because of rising temperatures. To a greater demand for sugars by the plants as a result of accelerated respiration. This is likely to decrease the price of the marketable sugar supply, thus decreasing the net income to farmers (Hatfield and Prueger, 2015). This process gets worse the longer it takes to fill the grain bin with grain or the fruit to ripen (Simpson, 2017).
- Bodies of water can get warm at night, causing a drop in yield (e.g., the soil for rice, and the water source for flowering) (e.g., beans).
- At a temperature between 20 and 30°C, photosynthesis begins to expand (Wahid et al., 2007). The yields productivity and plant populations are lower than diurnal from temps. which are usually between 34 to 30°C and 32°C (FAO, 2016)
- or seeds can be damaged at 37°, which is considered to be elevated temperatures, can experience irreversible damage if they are maintained above 30°C; different temperatures to accommodate to the fast changes in climate. Harm depends on how fast the

weather will change, how long it is done, and to what extent it has an effect (Wahid et al., 2007). The species, its maturity, and level of development all influence the outcome. It is expected that temperature has an impact on the frequency when the growth threshold for three staple crops – maize, rice, and wheat – will increase globally as the world gets warmer (Gourdji et al., 2013).

4.6.3. Changes in Precipitation Regimes

The overall range of precipitation is as well including variations in timing, frequency, and severity of particular occurrences, but there are far fewer changes in the actual duration and intensity of each rainfall are included in precipitation regime change.

Each of these factors plays a vital role in the yield of the crop. As temperature rises, crop evaporative demand will be significantly changes, so will changes in precipitation. There can be various types of moisture stress under crop growth, according to the point it is in which it occurs.

Most of the food, and since about 80% of the area grows crops that is fed by rainfall has the most critical food; about 60% of the global food supply is made possible by the shifts in precipitation regimes (Tubiello et al., 2007). How much precipitation a region gets and where it falls are important to farming will affect the production of crops. This applies to climate types around the world, as well as locations.

As climate change is taking place, it is likely that places will get higher amounts of rainfall and the further that those that were previously dry will see a reduction in it (Liu and Allan, 2013). Soil erosion and nutrient depletion can have a greater effect on soils in locations with less frequent periods of soil wetness during the year. Soils with lower organic carbon sequestration ability have a reduced water retention when the moisture content is low.

Furthermore, plants in nutrient-poor soil are slower to recover after receiving water than do plants grown in rich soils, which are particularly abundant in potassium (Lipiec et al., 2013). If climatic conditions are not predictable seasonal information to support growing their crops becomes obsolete, farmers would no

longer be able to rely on it for predictability. The changes in the seasons and the way nature works would increase difficulty for farmers to prepare and implement their crop production.

A later start to the rainy season is important because the later it increases the length of the plant's growth period, but also reduces the amount of time that plants have to produce yields, so eventually this decreases their growth (Linderholm,2005).

In photosensitive plants, prolonged changes in the rainy season period and adequate soil moisture that is, long, consistent rainy season lead to maturation of photosynthesis, resulting in reproduction cycle anomalies (e.g., reproductive failures). The other anticipated consequence of climate change is that the frequency of severe weather conditions will rise.

No drastic increase in the dryness or wetness is not necessarily needed to make the impact on crop yields more pronounced, but even in the worst cases, there are still major reductions in yield due to increased impacts of excessive weather conditions. Crops that are particularly susceptible to wet, heavy rain, hail, and floods may be damaged in the field. Because of the heavy rainfall, planting and harvesting in the field was often delayed by this season's rainy weather. There is little doubt that prolonged drought can result in total crop failure (Tubiello and van der Velde, 2010).

Figure 4.5. Bugs will eat More of our food with an increase in temperature.

Sources: Image by Flickr.

4.6.4. Pests

What happens in regard to pest-plants and pest activity is something called climate change, and it changes their interrelationships over time. The plants that have been affected by direct factors in their health by adverse weather conditions are also more susceptible to exposure to other conditions that lead to indirect ones.

The plants afflicted by water stress have fewer resistance to viruses, and if a condition is worsening, then they can compete with others who are increasingly losing water to monopolize the available resources (Simpson, 2017).

To expand on this, biocontrol can continue to be useful as long as it will still serves as a natural pest control in regions where natural enemies have not moved in yet, or the bio condition for the species is successfully assumed.

The appearance of pests is dependent on the surroundings' increase or decrease in temperature in need of habitats. It is now possible that because of the rapid increase in global warming, to that insects, whose body varies with the surrounding temperature, maybe moving into higher elevations (summer habitats) in search of additional locations (destination) (Bebber et al., 2013).

A pest response to changes in cropping trends in land use or climatic conditions may be regulated by changes in the methods used for their distribution. The cereal stem borers (Chilo, Sesamia, and Scirpophaga spp.), pod borers (Helicoverpa, Maruca, and Spodoptera spp.), aphids, and whiteflies are

major insect pests of cereals, pulses, vegetables, and fruit crops that may migrate to temperate regions (Sharma, 2014). The impact of pest epidemics will be distributed over a large geographical area; arthropod disease will dynamics; the activity and distribution of the arthropod population will vary; relationships of arthropods with herbivores will change; various kinds of herbicides could be ineffective, and others could outcompete arthropods; insect species will potentially go extinct; natural enemy distributions will vary; crops could be hard to control.

It is well accepted that weeds would be significantly impacted by climate change. The allowable temperatures for more invasive plants, including *Vallisneria* spp., cannot go north of 10 degrees warmer than those which are presently assigned because of the existing assignments, while those found to be unsuitable, cannot go any farther from their current location, as in the case of *Pistia stratiotes, Eichhornia crassipes* and *Salvinia auriculata* (Hussner et al., 2010).

Because these species have greater mobility, if not perfect, geographic, and climatic range, they would be more prevalent. Invasive species, such as *Imperata cylindrica* (cogon grass), *Pueraria lobata* (Kudzu), and *Striga asiatica* (witchweed), have traits that foster long-distance seed dispersal (Patterson, 1995a; Rejmanek, 1996; Dukes and Mooney, 2000; Malcom et al., 2002).

During winter, there is a spread in distribution of the Pueraria lobata as a result of a reduction in minimum temperatures to low −15°C (Ziska et al., 2010). This can also be done for the Striga, a parasite that develops in semiarid habitats at temperatures between 30°C and 35°C and propagates under the condition that soil fertility and cereal monoculture are poorly controlled.

A similar to that of increasing climate change in plant biology is said to have competition for resources with crops or weeds in distribution is said to play a role in environmental competition (Ziska and Reunion, 2007).

If it is accepted that hotter average temperatures promote Parthenium assimilation, it is also speculated that warmer temperatures can cause higher partitioning against root biomass. This species' greater root biomass helps to quickly and robustly regenerate trees after fuelwood loping (Kathiresan,2006). The competition among weed species and plants, and CO_2 levels can dramatically change crop productivity under various climate conditions (Patterson, 1993).

An average temperature rises of 3°C will, for example, promote perennial invasive C4 weed *Rottboellia cochinichinensis*, leading to substantial reductions in outputs in many critical crop systems (Patterson et al., 1979; Lencse and Griffin, 1991; Lejeune et al., 1994).

4.7. CONCLUSION

The emission of greenhouse gases has had much more of an effect on climate change than the conversion of forests to agriculture, which itself has played a role in emissions. The climate change is now threatening natural resources that which is vital to the food production, and on which agriculture relies. With regard to global warming, modest increases in yields may be possible, but net negative effects will become more prominent.

Rising global temperatures, more frequent floods, and worsening droughts can endanger farm production and rural livelihoods. Other types of ecological changes, environmental factors are more probable with climate change, which will include insect and parasitic plants as well as temperature change, would accelerate the range of various plant and animal diseases. The increasing demand for food and agricultural products such as soy, corn, would put more stress on the environment because of agricultural yields and populations are coupled with growth and profits.

In our day-to-day lives, these issues will be more or less likely minimized by technical advances and solutions to environmental destruction caused by agriculture. Policies and technologies that applied earlier have a lower potential for causing environmental damage than damage and for producing favorable outcomes and may thus have a slower expansion of both the number of negative and positive impacts.

REFERENCES

1. Hakeem, K. R. (2015). *Crop Production and Global Environmental Issues.*
2. (n.d). *Crop Production and Environmental Protection.* [e-Book] IPNI. Available at: http://www.ipni.net/publication/stewardship.nsf/0/1BEB3166687B91F285257BE50055366A/$FILE/StewSpec-EN-11.pdf (accessed on 7 July 2021).
3. Fao.org. (n.d). B1 - 1 Crop production and climate change | *Climate-Smart Agriculture Sourcebook* | Food and Agriculture Organization of the United Nations. [online] Available at: http://www.fao.org/climate-smart-agriculture-sourcebook/production-resources/module-b1-crops/chapter-b1-1/en/ (accessed on 7 July 2021).
4. FoodPrint, (n.d). *Pesticides in Our Food System* | FoodPrint. [online] Available at: https://foodprint.org/issues/pesticides/ (accessed on 7 July 2021).
5. Griffiths, H., (2003). *Effects of Air Pollution on Agricultural Crops.* [online] Omafra.gov.on.ca. Available at: http://www.omafra.gov.on.ca/english/crops/facts/01-015.htm (accessed on 7 July 2021).
6. Piri, I., Babayan, M., Tavassoli, A., & Javaher, M., (2011). The use of gamma irradiation in agriculture. *African Journal of Microbiology Research, 5*(32).
7. Shahid, M., Khalid, S., Abbas, G., & Shahid, N., (2015). [online] Research Gate. Available at: https://www.researchgate.net/publication/286922829_Heavy_metal_stress_and_crop_productivity (accessed on 7 July 2021).

Chapter 5

Progress and Challenges of Wheat Production

CONTENTS

In the chapter, Progress and Challenges of Wheat Production, several challenges that come in the way of maintaining wheat productivity is discussed. Some of the key challenges that are discussed are pests, diseases, and potential epidemics. While, emerging challenges such as new epidemics, Climate Change, and Other Long-Term Environmental Effects are also taken into consideration in this chapter. It also shed some light on the prospects of doubling global wheat yields. In addition, protecting yield potential is also discussed in this chapter. It also highlighted the importance of creating and exploiting new diversity. Finally, at the end, the significance of considering food safety in addressing challenges of wheat production is also highlighted.

5.1. INTRODUCTION

Cereals play an important role in satisfying the food demand of the growing population in the world, particularly in developing countries where the production system of cereals is the primary source of calorie intake and nutrition. The nutrition rich cereal is grown in diversified environments. Production of wheat occupies around 217 million hectares globally. Among all crops, it holds the position of highest acreage with an annual production of around 731 million tons. Wheat (*Triticum aestivum* L.) is one of the important staples of nearly 2.5 billion of world population and one of the principal cereal crops grown worldwide. Wheat is the major staple food crop that provides almost half of all calories in the region of West and Central Asia as well as North America. In terms of calories and dietary intake, being next to rice, wheat constitutes one of the key sources of protein in least developed countries and middle-income countries. The crop being cultivated in winter as well as spring in the world, winter wheat is grown in cold countries such as Australia, Europe, Russian Federation, the USA, etc., whereas spring wheat is grown in some parts of the USA and in Asia.

5.2. THE CHALLENGES OF MAINTAINING WHEAT PRODUCTIVITY: PESTS, DISEASES, AND POTENTIAL EPIDEMICS

With the increasing population in the world, food security projected to become more critical and increasing wheat yield potential in the developing world is at high priority. In order to achieve the objective, high-production irrigated areas will continue to play a major role.

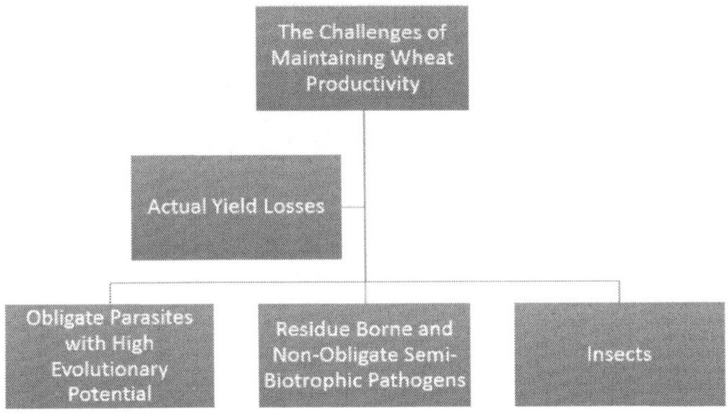

Figure 5.1. The Challenges of Maintaining Wheat Productivity.

According to Cook and Veseth (1991), absolute yields are determined by genetic potential, whereas the level of diseases, pests, as well as other threats determines actual yields. Gaunt (1995), stated that stand reducers, light reducers, tissue consumers, assimilate sappers as well as senescence accelerators are all ready to take advantage of the situation.

Therefore, getting aware of the pathogens, their distribution, ecology, virulence patterns and variability is important in reducing diseases and the gap between actual as well as attainable yields. Foliar diseases caused by obligate or semibio trophic pathogens with a high evolutionary rate (McDonald and Linde 2002) are the most important yield constraints in environments characterized by dense stands and high tiller density.

This is mainly due to the reason that some airborne parasites can migrate over long distances. In addition, they are more likely to overcome genetic resistance as compared to pathogen populations with a low evolutionary potential, which are confined to fields, i.e., soil-borne pathogens. In year 1996, Dubin stated that their direct effect on grain-fill, particularly after anthesis via reduced light interception and radiation use efficiency, logically results in lower yields.

5.2.1. Actual Yield Losses

Research papers on wheat diseases suggest that sizable areas of wheat are at risk to specific diseases or pests and frequently report impressively high potential yield losses. Generally, high yield loss figures are obtained in trials conducted on experiment stations or under controlled conditions, a commercial cultivar that has become susceptible often using as a check.

Although risks are not undermined and must be properly calculated, actual losses as well as crop health in fields of farmers are significantly different, as is generally the case in intensive production systems where modern and broadly adapted resistant cultivars are cultivated.

According to Oerke et al. (1994), the global average of actual yield losses caused by all wheat diseases was about 12.4% on an annual basis, including developed and developing countries. Unfortunately, precise data on actual yield losses caused by diseases in fields of farmers are difficult to assess or are often unavailable in developing countries.

A study covering 1976–2000 and including 18 diseases reported annual losses of 10–22% in Kansas (Bockus et al., 2001). Leaf rust caused by *Puccinia triticina* was by far the most important disease during this period, causing on average 3.48% losses in a year, followed by wheat streak mosaic virus (1.88%), and the Septoria complex (1.6%).

In the same study, barley yellow dwarf virus (BYDV) (causing 4.5% and 3.5% losses in year 1976 and 1987, respectively) or tan spot (causing 3.5% losses in year 1980 and 1987), all diseases taken individually caused less than 3% losses annually, and in general not more than 0.1–2% with the exception of leaf rust and outstanding outbreaks of *Septorias* (causing 5.8%, 3.5%, and 7.4% losses in year 1983, 1984, and 1995, respectively) (Bockus et al., 2001).

Due to leaf rust of around 3.7%, an estimate calculated on 22 developing countries growing more than 100,000 ha of wheat indicated an area-weighted average yield loss over 10 years (Marasas et al., 2004) which is in agreement with above data. In the Indo-Gangetic Plains, it has been observed that spot blotch caused by *Cochliobolus sativus* is reported to prevail on about 9 million hectares of wheat grown after rice and yield losses of 20% (Duveiller 2004).

However, depending on the level and duration of dew as well as heat stress, most likely only 40% of this area is actually suffering losses in the range of 5–15% every year (Sharma and Duveiller 2004). It is important to note that a disease only takes place in case inoculum is present, the environment (soil, climate, and cropping system) is favourable, and the host is susceptible.

In addition, the relationship between diseased leaf area or any other disease scoring scale and yield is not straightforward due to the reason that the physiological implication of symptom assessment and effect on yield is not always proportional or linear.

The extent of damage may vary significantly as per the host physiological growth stage at the onset of an epidemic. Moreover, it often depends more on the duration of the green leaf area as compared to the amount of disease area estimated with the help of a visual scale.

5.2.2. Obligate Parasites with High Evolutionary Potential

5.2.2.1. The Wheat Rusts

According to Saari and Prescott (1985), the rust diseases of wheat have historically been one of the major biotic production constraints worldwide. Since the semidwarf spring wheat of the Green Revolution came to occupy most of South and West Asia in the 1960 s, stem or black rust caused by *Puccinia graminis* has been under control.

Particularly in Asia, stripe (or yellow) rust (Chen 2005) caused by *Puccinia striiformis* and leaf (or brown) rust caused by *P. triticina* continue to pose a major threat to wheat production over large areas. In this continent, stripe rust as well as leaf could affect production on approximately 43 (46%) and 60 (63%) million hectares, respectively, in case susceptible cultivars are grown (Singh et al., 2005).

Rust fungi only survive on living plants and are obligate parasites. Irrigated agriculture, high input and the existence of a cool off-season location where accessory hosts or wheat is found, promote the carryover of inoculum between seasons.

Now, a number of races are known to occur worldwide. The presence of races that are capable of overcoming various resistance genes as well as their combinations has been demonstrated for all three rust fungi.

New races may take place with the mutation, sexual recombination (not known for *P. striiformis*) or somatic hybridization followed by selection in case a new race has a selective advantage (Singh et al., 2002).

The strategies to limit damages caused by wheat rusts followed the principles of a coordinated anticipatory breeding approach, especially in the last decades (McIntosh and Brown 1997; Singh and Rajaram 2002; Singh et al., 2002).

The methodology has been successful, particularly against yellow rusts as well as leaf and is summarized hereafter. Pathogen monitoring through scouting of commercial fields and observing trap nurseries at favorable locations or relevant hot spot along with limited sampling for race analysis confirms the prevalence of major existing races and allows the early detection of new races. Enhanced information related to the effectiveness and genetic basis of resistance in important wheat cultivars is obtained together with the testing of new wheat germplasm developed by National Agricultural Research Systems (NARS), Advanced Research Institutes (ARIs) as well as International Agricultural Research Centres (IARCs) at relevant hot spot sites. Diverse resistance sources are used for the enhancement of germplasm. Breeding for durable resistance based on the accumulation of additive minor genes is emphasized through the use of race nonspecific (slow rusting) resistance. According to the views of Caldwell (1968), slow rusting is such type of resistance where diseases progress at a retarded rate.

Consequently, intermediate has to reduce disease levels against all path types of a pathogen. Adult plant resistance characterizes it despite compatible infection type at seedling stage. DNA marker-assisted selection (MAS) should be used when it is found to be feasible in the process.

Through meaningful and efficient seed production programs, seed multiplication and distribution of diverse resistant improved genotypes are to be promoted.

However, schemes related to participatory variety selection may prove useful in rapidly replacing old varieties in fields of farmers. In rust-prone areas, only rust resistant cultivars should be recommended as a rule. Lastly, resources for training, research as well as infrastructure are of primary importance.

5.2.2.2. Powdery Mildew

Powdery mildew is an important disease of wheat worldwide. It is caused by *Blumeria graminis* f. sp. *Tritici*. It is found especially in highly productive areas with a maritime or semi continental climate (Cunfer 2002).

Powdery mildew conidia do not spread over very long distances; however, the evolutionary rate of fungus is high as compared to rusts. Powdery mildew is important in China as well as in South America, where at least 12 million hectares can be affected (Costamilan 2005).

It may also be cause for concern in western India as well as Pakistan, where susceptible varieties are grown and the weather is favorable. In year 2005, Wang stated that since natural epidemics of powdery mildew do not take place in Mexico where CIMMYT has its main breeding operations, most of the breeding materials sent from Mexico to the regions are susceptible when grown in areas prone to the disease.

The most economical and environmentally safe means of controlling powdery mildew is the utilization of resistant cultivars.

Knowing the virulence pattern of isolates collected over large areas may prove useful at the time of combining race-specific resistance genes (Wang et al., 2005). However, this approach is not sufficient since race-specific genes result in a rapid build-up of the pathogen population possessing matching virulence genes.

To find adult plant resistance such as that in North America's cultivar Knox 62 together with its derivatives that have shown durable resistance over a long period is a more sustainable objective. In year 2006, Lillemo et al., stated that the bread wheat cultivar Saar exhibited a high level of this type of resistance in the fields. In addition, at least three genes controlled it.

Lillemo et al. (2005) also found that genes Lr46/Yr29 and Lr34/Yr18 contributed towards reducing the levels of powdery mildew. Understanding the reaction mechanism against obligated parasitic fungi controlled through genes against a number of diseases like yellow rust as well as the leaf and powdery mildew shows a new research challenge to unravel the basis of race nonspecific adult plant resistance against obligated parasites.

5.2.3. Residue Borne and Non-Obligate Semi-Biotrophic Pathogens

5.2.3.1. Tan Spot

Tan spot is found in South and North America, Europe, Central and South Asia. It caused by *Pyrenophora triticirepentis* and has a wide distribution (Friesen et al., 2005). According to Bockus and Shroyer (1998), the disease is seed-borne as well as favored in wheat grown after wheat under zero tillage. The fungus exists on residues of crops at the surface of soil where the sexual stage (pseudothecia containing ascospores) begins.

In year 1998, Annone observed that tan spot incidence and severity has been linked with the change in the variety composition and expansion of zero tillage cultivation, especially in Paraguay, Argentina as well as Brazil where it is considered as major leaf blight responsible for potential yield losses of 10–20%. Tan spot is common in Central Asia where mono-cropping is common and where long winters do not allow fast stubble decomposition.

Damages are the outcomes of destruction of the flag leaf, so the plant is robbed of assimilates or photosynthesis products otherwise destined to fill the grain. Tan spot is capable of killing the leaves after heading, which results in premature death of the infected plants.

5.2.4. Insects

There are a number of insects that have been described on wheat worldwide. Most of these insects cause damage that is either insignificant or limited to isolated areas. However, other pests inflict serious yield and forage losses. According to Miller and Pike (2002), some of the problems of these pests are directly linked to the unique farming system that is employed in a specific area, whereas other pests are opportunistic or generalist herbivores that do not particularly target wheat as a host. Usually, feeding as well as chewing insects does not cause major direct damages in wheat unless populations reach very high levels; infestation by several aphids feeding on wheat result in higher BYDV incidence, such as *Rhopalosiphum padi* or *Sitobion avenae*. In some areas, Russian wheat aphids (*Diuraphis noxia*) as well as greenbug (*Shizaphis graminum*) cause damage by injecting a toxin while feeding on leaves. Hessian fly (*Mayetiola* destructor) may cause significant damages in North Africa, the Mediterranean region, and parts of West Asia. It causes kernels to stop filling, tillers to stop growing, or stems to lodge. In year 2002, Miller and Pike stated that sources of host plant resistance have been identified in cultivated as well as wild wheat and they are mostly used in breeding programs.Suni bug (*Eurigaster integricepts*) is one of the most important wheat insect pests found in some developing countries, particularly in dryland areas in North Africa and West Asia. It injects a toxin into the grain that results in both shriveled grain and inferior dough making properties for bread. cereal leaf beetle *Ulema melanopus* causes severe crop damage by reducing the photosynthetic area of plants in Tajikistan and Afghanistan.

5.3. EMERGING CHALLENGES: NEW EPIDEMICS, CLIMATE CHANGE, AND OTHER LONG-TERM ENVIRONMENTAL EFFECTS

Changes in cropping systems that are outcomes of changes in rotations or tillage practices influence the prevalence as well as survival of residue borne pathogens. In case new agronomic methods are not accompanied by the adoption of cultivars with adequate resistance, they can lead to a shift in the disease spectrum.

Figure 5.2. Continuous bed planting for wheat.

Source: Image by nue.okstate.edu.

The occurrence of a "new" disease in a determined geographic area or cropping system is rare, with the exception of an accidental introduction of a new race or parasite. However, a green bridge between two crops may arise, and a disease that was earlier considered as an oddity may become an important constraint in case a minor pathogen is present and remains marginal because of an unsuitable environment, but climatic conditions become favourable suddenly for its development.

Wheat blast caused by *Magnaporthe grisea* is now a disease and in need of urgent attention by wheat breeding programs in Brazil after initial reports of its occurrence in the mid-1980 s (Igarashi et al., 1986). The disease poses a potential threat to wheat due to the lack of resistance in other geographical areas with similar climate.

Yet it is also a potential threat to these crops due to the host range maize and triticale are susceptible (Urashima et al., 2005). It is observed from recent studies that wheat blast did not originate from the rice blast fungus and strains of wheat as well as rice were genetically dissimilar (Urashima et al., 2004, 2005).

According to Urashima (2005), probably the fungus that attack wheat is an adaptation to wheat as its new host of blast isolates from a weed (*Digitaria insularis*). It may completely hamper the grain fill and attack at the rachis level that causes total losses in susceptible genotypes, depending on the timing of infection. Although current economic losses due to *Rhizoctonia cerealis* are generally low, in some regions with intensive wheat cropping systems, such as parts of China, sharp eyespot caused by this disease may be an emerging problem. Recently, sharp eyespot was diagnosed in Egypt where the incidence was approximately 20–30% of sampled areas (Hammouda 2003).

Change of climate may have a long-term effect on wheat-growing areas; it could be inducing a shift in the geographical area grown to wheat as well as associated biotic constraints to higher latitudes. In the same way, current pathogen as well as pest spectrum may evolve in a given wheat area.

For instance, as a result of possible climate shifts in the Indo-Gangetic Plains—currently part of the favorable, high potential, low rainfall mega-environment, irrigated, which accounts for 15% of global wheat production up to 51% of its area might be reclassified as an irrigated, heat-stressed, short-season production mega-environment by 2050 (Ortiz pers. commun.).

An increase in spot blotch severity and incidence can be anticipated in optimum wheat-growing areas where the disease does not figure prominently today, in case this scenario proves to be correct.

5.4. PROSPECTS OF DOUBLING GLOBAL WHEAT YIELDS

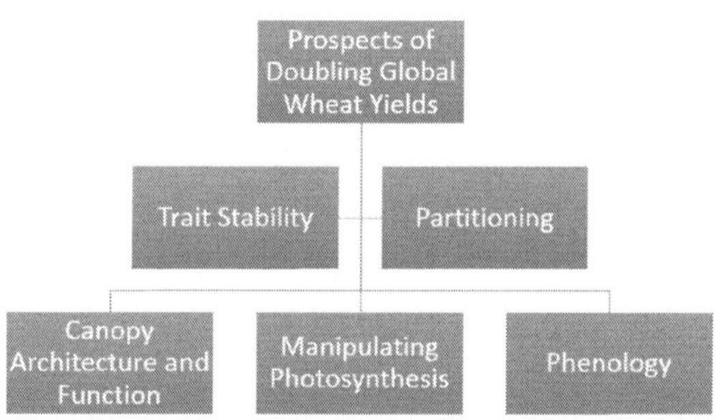

Figure 5.3. Prospects of doubling global wheat yields.

In spite of the fact that an adequate supply of food can be achieved at present for the current global population, sustaining this into the future will be difficult in the face of a steadily increasing population, increasing wealth and a diminishing availability of water together with the fertile land for agriculture.

This problem gets compounds through the new uses of products related to agriculture, such as biofuels. Wheat alone provides ≥ 20 percentage of the protein together with the calories for a population of the world and the value as well as need to increase the production is widely recognized.

At present, the average wheat yield of the world is around 3 t/ha. However, there is considerable variation among countries, with region-specific factors limiting yield, each requiring solutions individually. A complex challenge is to deliver increased yields in any situation that is unlikely to be solved by single approaches, and a multidisciplinary integrated approach to crop improvement is required. There are three specific major challenges and that are protecting yield potential, increasing yield potential, and increasing resource use efficiency to ensure sustainability. In many countries, yields at the farm gate have stagnated or are increasing at less than half the rate required to meet the projected demand since the green revolution. Large gains can still be achieved by improvements in agronomy in some countries, but the yield gains will only be achieved by further genetic improvement in many other countries. The importance as well as the prospects for upgrading individual traits is presented. Therefore, we can say that there are opportunities for an increase in yield, but a major challenge will be avoiding a simultaneous increase in resource requirements.

5.4.1. Specific Traits

5.4.1.1. Trait Stability

Year to year variability in yields can be significant in many regions. Farmers often prefer guaranteed minimum productivity instead of gambling on high yields, with the alternative being very poor yields. Production stability together with the consequent influence on markets is also preferred on a global scale.

Therefore, trait stability is a prerequisite for high-yielding varieties, particularly yield but also quality attributes. Traits require being robust across a range of environments and on a year-to-year basis.

5.4.1.2. Canopy Architecture and Function

Production can be increased by the establishment of a photosynthetically active canopy, optimized for the production of photoassimilates that determines the final yield of grain. Approaches to increasing carbon fixation are optimizing canopy architecture and light capture efficiency (maximizing leaf angle for light interception or targeting complete canopy closure), extension of the grain-filling period and modification of intrinsic photosynthetic efficiency. Clearly, traits to avoid disease spread and lodging need to be considered. Moreover, architecture needs to be optimized for maximum light capture. A secondary but important role of the canopy is N - storage (as protein) as well as N - assimilation, and the subsequent effective remobilization of this N to the grain at the time of crop maturation.

5.4.1.3. Manipulating Photosynthesis

The primary determinant of crop biomass is the cumulative photosynthesis of the growing season. With the condition that other constraints do not become limiting, increasing photosynthesis, for instance, by expanding the substrate CO_2, as in free-air CO_2 enrichment experiments, has been demonstrated clearly to increase yields (Long and Ainsworth 2005).

The cumulative photosynthesis can be increased through increasing light interception, its duration or photosynthetic rate. In some regions, it may be possible to increase the duration of photosynthesis, but the growing season is most often constrained either by the cropping system or by environmental factors (water availability and low/high temperatures).

However, still opportunities are there to increase photosynthesis by manipulating senescence to delay its onset and by improving early vigor. Light interception is very efficient (Horton 2000) despite the fact that there is considerable variation in the structure of modern wheat canopies (for instance, flag leaf size and leaf angle). Potential increases in photosynthesis are small, while further improvements in canopy architecture may be possible (Murchie et al., 2009 and Reynolds et al., 2012).

The biggest potential gains in cumulated photosynthesis may become possible to achieve by increasing the photosynthetic rate. In case of wheat, only 4.6% of the intercepted radiation is converted to photosynthetic. Moreover, there is clearly a good deal of room for improvement (Zhu et al., 2010). In year 2011, Parry stated that a number of potential ways to increase photosynthetic rate have been identified. Many of them aim to focus on expanding the concentration of CO_2 within the leaf.

According to Fischer (1998), simply increasing stomata as well as mesophyll conductance will increase photosynthetic rate and yield however, it may decrease water use efficiency. With the introduction of a variety of CO_2 concentrating mechanisms, internal CO_2 concentrations could be increased. Although worthwhile, each requires a number of technological hurdles to be overcome (Hibberd et al., 2008; Zarzycki et al., 2012; Price et al., 2013). A simpler approach and perhaps an alternative are to replace the CO_2-fixing enzyme, Rubisco, with one that would deliver higher photosynthetic rates. Isolated from different species, there is variation in the kinetic properties of Rubisco (Parry et al., 1989; Delgado et al., 1995; Galmés et al., 2005) that is sufficient to at least in theory confer superior characteristics to photosynthesis in wheat (Zhu et al., 2010 and Parry et al., 2011). However, at present it is not possible to do this in wheat (Parry et al., 2012).

An alternative approach would be to have an extended period of grain filling. For instance, an optimized canopy might involve early and complete canopy closure together with early flowering. In the same way, because of delayed senescence, an extended grain-filling period may be beneficial. However, extending the grain-filling period is fraught with difficulties and may expose sensitive tissues to late frosts or nonoptimal water supply or extremes of temperature.

Excess cool as well as damp conditions at the time of crop maturation while positively influencing yield may substantially compromise with quality even in temperate regions. According to Derkx (2012), delaying senescence may expand carbon fixation but may limit N-remobilization to grain tissues, impact on quality or even sink capacity for assimilation of carbon.

5.4.1.4. Phenology

Breeders have adapted the length of the crop cycle to each particular environment to avoid climatic restrictions such as high temperatures at the time of grain-filling period and frost during flowering, which could cause irreversible damage to crop yield for years. Therefore, the length of the total cycle is often well balanced for a specific environment (Isidro et al., 2011).

Despite the fact that components of yield are built throughout the phases of the crop cycle, some phases are more critical in terms of grain yield definition as compared to others. In this context, in barley as well as wheat, the conditions during the period immediately preflowering is important for the determination of yield due to the fact that the most important yield components are defined in that period. According to

Fisher (1984), the grains per spike and number of tillers per unit area that are the main components of grains per unit area are established in the period between 30 days before and 10 days after flowering.

Actually, that period was named the "critical period" for determination of yield. In addition, the number of fertile florets, the main component of grains per spike, is defined during the critical period. In year 1998, Miralles stated that as floret initiation is not the bottleneck determining the final number of grains per spike, as more than 10 florets primordial are initiated into the spikelet, the aim is to promote survival of floret.

Spike and stem compete for assimilates at the time of the critical period, therefore, the higher assimilate partitioned to the spikes, the larger the number of fertile florets at flowering, and as a consequence of that the final number of grains are increased (Fischer 1984).

Without changing the length of the cycle, other studies proposed an extension of the critical period as a strategy to reduce the stem-spike competition and promote floret survival due to heavier spikes at flowering (Halloran and Pennell 1982; Slafer et al., 1996).

As per studies conducted by Miralles (2000) and Gonzalez (2005) under controlled and field conditions, it is confirmed that an extension of the period from terminal spikelet formation to the onset of flowering promoted spike fertility, increasing the number of grains per unit area.

The Ecophysiological basis of the increase in spike fertility due to an elongation of the critical phase is associated with (i) a delay of the beginning of spike growth when floret primordia start to die, (ii) less competition between stem and spike determining heavier spikes at flowering, and (iii) allowing a sustained floret development because of a reduced inter floret competition within the spikelet allowing more floret primordial in order to reach the fertile floret stage (Gonzalez et al., 2011).

5.4.1.5. Partitioning

Especially under high yield potential conditions, the green revolution was based on plant height reduction to avoid lodging and allowing higher inputs to be applied to the crop (e.g., fertilizers) to increase yield (Fischer and Stockman 1986).

Nevertheless, the most revolutionary effect of the green revolution was a consistent and clear development in the internal partitioning of the plants, promoting increases in spike weight because of a reduced competition between stem as well as spikes for carbohydrates. Actually, the spike stem ratio was increased at a rapid pace in new cultivars that were released after the green revolution as compared to older cultivars (Calderini et al., 1997).

HI was increased because of plant height reduction (that is, the biomass partitioning to the reproductive organs, the grains). As reported for U.K. cultivars, the HI of modern wheat genotypes is 55% (Foulkes et al., 2009b). This value is close to the theoretical upper limit calculated by Austin (1980) that is 62%.

Therefore, in this country, few opportunities are anticipated to continue increasing grain yield through higher partitioning. On the other hand, although values of 49% have been recorded in Hebei Province (Zhou et al., 2007), the reported values of HI reached by modern Chinese wheat cultivars are between 42% (Xiao et al., 2012) and 46% (Zhou et al., 2007).

There is an opportunity to improve this trait beyond doubt. However, further reduction in height of plant is unlikely to be a useful strategy for breeders in the future, as the relationship between yield and plant height has an optimum (Richards 1992; Miralles and Slafer 1995).

Actually, plants height under 0.70 m or above 1 m result in the production of a lower biomass as well as lower partitioning to reproductive organs, respectively.

In Northern China, most of modern wheat is already within the range of the optimum response of grain yield to plant height, that is, 0.7 m–0.9 m (Zhou et al., 2007 and Zheng et al., 2011). In the same way, in addition to biomass production (see above), the strategies of focusing on increasing yield components per se (grain number mentioned above or grain weight in the following section) could be more helpful for improving yield potential than attempting plant height and biomass partitioning as in the past.

5.5. PROTECTING YIELD POTENTIAL

Abiotic stress: what is threatening yield potential, insufficient genetic advances or climate change?

Stagnation in yield for wheat, bread as well as other cereals has occurred during the last decades in China (You et al., 2009; Piao et al., 2010) as well as in most European countries (Peltonen-Sainio et al., 2009; Brisson et al., 2010; Olesen et al., 2011) and also in other parts of the world (Calderini and Slafer 1998; Ladha et al., 2003).

The case of France illustrates how change in climate challenges even the highly advanced agriculture of a developed country. Since the middle of the 1990 s, a stagnation of potential bread wheat yield has been observed in

France. Therefore, the possibility of a decline in breeding progress has been suggested (Brisson et al., 2010).

However, since the end of the 1980 s, a recent study has shown that the adverse effects of climate change have counterbalanced the genetic process totally or partly (Oury et al., 2012). Thus, in some degree, yield potential trials may be influenced through adverse climatic conditions, the major factor being climate warming (Oury et al., 2012).

Therefore, during the last decades, it may be that genetic progress has not experienced any weakening and may be considered as continuous (Oury et al., 2012). In recent years, in the study of Peltonen-Sainio et al. (2009), this linear trend has also been found in Finland with no perceptible limitation to genetic progress.

An increase in mean temperature may negatively affect yield physiology at other levels apart from the direct effects of a higher temperature increasing photorespiration and dark respiration losses or shortening the crop cycle. For instance, it may have a direct or indirect effect increasing frequencies for spring as well as winter droughts (Brisson et al., 2010).

More precisely, drought and/or high temperatures during filling of grain lead to reduced grain weight (Gooding et al., 2003) and drought at the time of stem elongation results in reduction in vegetative dry matter with a concomitant decrease in a number of grains per unit area (Jamieson as well as Sinclair 2006).

Whatever the mechanism, estimated reduction in productivity per 1°C increase in temperature at the time of growing season may reach 5% or more of actual yields easily (Kalra et al., 2008; You et al., 2009; Gallais et al., 2010).

In other parts of the world, the adverse effect of global warming on crop productivity has already been demonstrated. For instance, in Asia, the increase in night temperatures has been related to the stagnation or the decrease in rice yields (Ladha et al., 2003; Peng et al., 2004).

In the same way, the harmful effects of high temperatures were identified for soybean as well as wheat yields in the United States of America (Lobell and Asner 2003), and for wheat yield in India (Kalra et al., 2008) and China (You et al., 2009). In addition, some of the above studies (Kalra et al., 2008; Oury et al., 2012) mark the importance of the fragility of intensive agriculture that appears even more sensitive to climate when yields are high.

In the case of China, for instance, present knowledge and understanding does not allow a clear assessment of the impact of anthropogenic climate change on agriculture and water resources of China. As per the views of Piao (2010), future study must improve regional climate simulations, especially of precipitation, and develop a better understanding of the managed as well as unmanaged responses of crops to changes in pests, diseases, climate, and atmospheric constituents.

In addition, it is often assumed that increasing social concerns related to pollution and environmental sustainability of agriculture in China will further trigger the research interest for a higher efficiency in the use of resources by crops (Fan et al., 2012).

A 4.5% reduction in wheat yields is attributed to rising temperatures over the period 1979–2000 for China, countrywide (You et al., 2009). Warmer daytime temperatures are likely to have decreased yields of wheat over a wide range from 6% to 20% per °C (Tao et al., 2008), which stresses the need for crop-specific and regional studies.

Therefore, if we aim to cope with the reductions in crop yield caused by climate change, increasing the genetic gains beyond the current values is a requirement.

5.5.1. Biotic stress

Pathogens (bacteria, oomycetes, fungi, viruses, and nematodes) are among the major biotic stresses affecting plants, affecting crop production principally by decreasing quality as well as yield. In year 2005, Strange and Scott observed that conservative estimates of global losses to plant pathogens across all plant industries are in the order of 10%.

It is very difficult to estimate the overall global impact of diseases in wheat production. A number of published accounts exist of the impact on production of regional epidemics of a single disease and of the impact of individual diseases on yield of a susceptible genotype.

For instance, in susceptible cultivars, stripe rust of wheat can cause yield losses in excess of 70%. In most wheat growing regions of the world, sporadic epidemics of the disease have caused substantial losses. Stripe rust epidemics in selections from the CIMMYT generated wheat line caused significant yield losses in Ethiopia, Yemen and Iran in year 1993 (McIntosh et al., 1995).

In China, winter wheat production is also affected by recurring epidemics of stripe rust (Stubbs 1985) with epidemics in 1950, 1964, and 1990 estimated to have incurred losses of 6, 3, as well as 2.5 million tons (Wan et al., 2007). In Australia, wheat yields are comparatively low mainly because of lack of water. A recent analysis of the impact of diseases in wheat production conducted by Brennan and Murray (2009) projected that they cause an existing average annual loss of AUD$913 million or $76.64 per hectare to the Australian industry of wheat. Over the past decade, the expected losses equated to about 19.5% of the average annual value of the wheat crop with five diseases accounting for about 65% of the overall loss.

Those five diseases were yellow spot (23% of overall loss that was caused by Pyrenophora tritici - repentis), stripe rust (14%; Puccinia striiformis f. sp. tritici), Septoria nodorum blotch (12%), crown rot (8% that was caused by Fusarium), as well as root-lesion nematode (8%; Pratylenchus neglectus).

The yield losses due to Fusarium head blight (FHB) is another combined example from China. FHB is the most important disease of wheat production that affects the North-east spring wheat and Southern winter wheat growing areas in China.

It is important to note that these are the largest wheat producing areas in China (Lu and Chen 2012). Global warming is resulting in the early onset of the rainy season and combined with changes in farming systems, specifically with growing residues of corn stubble resulted in a heavy epidemic of FHB in year 2012 and as a result, it caused serious yield losses in these regions (Zhang et al., 2012).

In this case, the lack of both FHB-resistant resources as well as FHB resistance cultivars adapted to the winter wheat growing region are the core restrictions of protecting yield potential in these areas.

Obviously, protecting production of future from losses of the magnitude projected in the study conducted by Brennan as well as Murray (2009) will be dangerous in case wheat production of world is to double. In this present era, chemical control is used in various parts of the world for controlling disease in broadacre cropping comprising wheat.

For instance, in eastern Australia, approximately $280 million has been spent on fungicides in order to control stripe rust in wheat following a single incursion of the wheat stripe rust pathogen from North America in year 2002 (Wellings, 2007). Nevertheless, concerns over the use of chemicals in the production of food will increase the role of genetics in controlling diseases.

5.6. CREATING AND EXPLOITING NEW DIVERSITY

Figure 5.4. Engaging in cooperation for exploiting diversity of wheat.

Source: Image by ElDiario.ec.

The identification as well as exploitation of genetic diversity for the varieties of key traits defined above is one of the fey features of conventional breeding. A chief concern for cereal breeding programs is that the disparity in such key traits has gradually become more restricted, at least between selected breeding germplasm for maize.

In year 2008, Mujeeb-Kazi stated that additional variation in such traits has been generated, for instance, by wide crossing and by the recreation of synthetic wheat. A number of commercial cultivars with FHB resistance and anticipated yield potential have been reported in Henan Province, China (Yang et al., 2009; Zhang et al., 2012).

These have been derived from common wheat, its wild relatives as well as various bridge materials, for example, perennial *Elytrigia elongate*, *Triticale*, *Leymus*, perennial *Elytrigia intermedia*, etc., by multistep conventional wide crossing assisted by techniques related to tissue culture.

This exhibits the potential importance of creating as well as exploiting fresh diversity by wide crossing to increase the common wheat gene pool for expanding wheat yield in the forthcoming period. In order to generate additional variation, mutagenesis can also be used.

As per the views of Parry (2009), worldwide more than 1000 wheat cultivars derived from mutagenesis have been commercially released. Genetic transformation also produces additional variation, although lines with interesting traits (e.g., drought tolerance, pest resistance, quality) have been evaluated, and few have yet been commercialized.

5.7. THE IMPORTANCE OF CONSIDERING FOOD SAFETY IN ADDRESSING CHALLENGES OF WHEAT PRODUCTION

Figure 5.5. Considering food safety in wheat production.

Source: Image by CBS Baltimore.

The definition of food security that is used most commonly is given by FAO of UN: 'Food security takes place when all people, at all times, have economic, social and physical access to safe, nutritious and sufficient food to meet their food preferences and dietary needs for an active as well as healthy life.' In simple terms, food is required to be nontoxic to eat as well as healthy and accessible.

Western consumers might put the same in comparatively more strong way, with several periods of ample food leading to food security escalating above accessibility as a concern. It is important for plant breeders to be aware of this.

For the reason that not only consumer safety and security must be given significance, but also because the food industry is tremendously sensitive to the risk of poor publicity for its products and companies will act promptly in case substituting to a different supplier or raw material decreases the probability of a product attracting negative titles.

The creation of the processing contaminant acrylamide in the course of the high-temperature cooking in addition to processing of foods made from cereals, potatoes as well as other crops is an example of just such a problem related to food safety and one that the food industry would absolutely not want being made any worse.

In year 2002, acrylamide that was discovered in a range of mainly plant-derived foods (Tareke et al., 2002) is categorized by the International Agency for Research on Cancer as a Group 2a carcinogen ('probably carcinogenic to humans') (International Agency for Research on Cancer, 1994) and the World Health Organization. As per the views of Friedman (2003), it also has effects on reproductive as well as neurological systems at high doses.

Acrylamide takes place because of the Maillard reaction. It is an umbrella name that is given to a series of nonenzyme reactions among sugars and amino groups, primarily those of amino acid that occur during high-temperature cooking and processing, involving baking, roasting as well as frying. Thorough explanations of the Maillard reaction have been provided earlier by Nursten, 2005; Mottram, 2007; Halford et al., 2011.

It was considered as an important reaction for the food industry long before acrylamide was found to be one of its products for the reason that it also produces the melanoidin compounds that give baked, fried and roasted products their color, and a host of other compounds that impart flavor as well as aroma (Mottram, 2007; Halford et al., 2011).

Due to this, the situation becomes more difficult for the food industry since these are the compounds that give foods the aromas, texture, flavors, and colors that define brands and are demanded by users.

The Maillard reaction have need of a reducing sugar like fructose, glucose or maltose, even though sucrose will participate in case it is first hydrolyzed through thermal, enzymatic, or acid-catalyzed reaction (De Vleeschouwer et al., 2009).

The condensation of the carbonyl (C=O) group of the decreasing sugar with an amino group is the first step. This creates a Schiff base that cyclizes and then undergoes rearrangement, enolization, dehydration, fragmentation as well as deamination in order to give rise to sugar dehydration and fragmentation products containing one or more carbonyl groups.

These carbonyl compounds react with amino groups in addition to other components that result in the formation of a plethora of diverse compounds. One of the important reactions is Strecker degradation, whereby an amino acid is decarboxylase as well as delaminated to give an aldehyde, and a Strecker-type reaction including asparagine is the chief route for formation of acrylamide (Mottram et al., 2002; Stadler et al., 2002; Zyzak et al., 2012).

As a matter of priority, the WHO/FAO Joint Expert Committee of Food Additives (JECFA) in addition to other risk assessment bodies have suggested that acrylamide levels in food should be reduced (Joint FAO/ WHO Expert Committee of Food Additives, 2011).

Moreover, in early 2011, the European Commission issued 'indicative' levels for acrylamide in food (European Commission, 2011). Indicative values are not considered as safe thresholds. However, they are proposed to specify the requirement for an investigation into why the level has been surpassed. It is important to note that products that have been found to surpass the indicative value have fascinated negative headlines.

Some of the most important contributors to dietary acrylamide intake across Europe (European Food Safety Authority, 2011) are wheat products together with coffee as well as potato products. In order to reduce acrylamide formation by modifying food processing, manufacturers have devised many strategies, and these have been compiled in a 'Toolbox' produced by Food Drink Europe (2011).

Additional development may become possible through variety selection as well as optimizing crop management (reviewed by Halford et al., 2012). The concentration of free asparagine is the major determinant of acrylamide-forming potential in wheat flour (Muttucumaru et al., 2006; Granvogl et al., 2007; Curtis et al., 2009) and the same is true in case of rye (Curtis et al., 2010; Postles et al., 2013) and probably for other cereals as well.

Therefore, decreasing the free asparagine concentration of wheat grain is a main concern. Moreover, there should be no doubt that manufacturers of food believe plant breeders to involve on the acrylamide issue.

While selecting for yield or other traits, breeders must also beware of inadvertently increasing acrylamide-forming potential for the reason that there is evidence from numerous crop species of links between asparagine metabolism, yield as well as protein content.

For instance, simultaneous silencing of asparagine synthetase genes *AST1* and *AST2* in potato tubers resulted in small tubers in field-grown potatoes that were done in order to decrease acrylamide-forming potential (Chawla et al., 2012). By silencing of *AST1* on its own, the problem was overcome, which as a result reduced asparagine levels in tubers without reducing tuber size.

This was possible because *AST1* controls asparagine levels in tubers whereas *AST2* has a wider role, however, the link between asparagine metabolism as well as yield is clear. In year 2012, Pandurangan conducted a study on soybean that exhibited a positive correlation between free asparagine levels in developing seeds as well as protein concentration at the time of maturity.

In year 2010, Gaufichon listed the biological functions of asparagine synthetase in plants and that are metabolic nitrogen recycling and flow in vegetative organs; nitrogen mobilization in the course of seed germination; nitrogen remobilization from source organs in order to sink organs and seed nitrogen filling as well as maturation.

In year 2007, Lea reviewed asparagine in plants more broadly by highlighting the accumulation of free asparagine at the time of normal physiological processes like nitrogen transport and seed germination also in response to a diversity of stresses, together with drought, mineral deficiencies, toxic metals, salt and pathogen attack.

It was concluded by the authors that now there is a clear indication that free asparagine accumulates in plants during times of low rates of protein synthesis joined with an abundant supply of condensed nitrogen. These studies display that to develop yield modification of wheat, physiology could inadvertently affect asparagine concentration in the grain, and this is something that breeders must be aware of.

Acrylamide is not only processing contaminant that causes concern. Furan (a heterocyclic compound made up of a five-membered aromatic ring involving four carbon atoms as well as one oxygen) and more complex compounds that comprise furan rings are also becoming the center of focus.

Furan is classified by the International Agency for Research on Cancer as ‘possibly carcinogenic’ to humans (Group 2B) (International Agency for Research on Cancer, 1995) and lays the foundation for liver cancer in rodents (Leopardi et al., 2010).

Furan forms through the Maillard reaction just like acrylamide; nevertheless, it also rises from the thermal degradation of polyunsaturated fatty acids, directly from ascorbic acid and from sugar breakdown (Perez Locas & Yaylayan, 2004; Crews & Castle, 2007; Moro et al., 2012). Hydroxymethylfurfural (HMF) is a compound that has attracted some attention in bakery products (Ramírez-Jiménez et al., 2000) that comprises of a furan ring with both aldehyde as well as alcohol functional groups.

5-sulfoxymethylfurfural is one of the metabolic products of HMF that can form adducts with DNA or proteins. It has also shown potential toxicity together with carcinogenicity in studies of rodent toxicology (Husøy et al., 2008).

While breeding for yield or other traits, it is important that breeders take on board the need to consider the potential for processing contaminant formation. Contaminants occur from other sources, obviously with mycotoxins formed by fungal activity being a noticeable instance.

Nevertheless, it is extremely doubtful that developments in yield would bring an increase in mycotoxin contamination as well as fungal infection.

5.8. CONCLUSION

At the end, it is concluded that there are plethora of opportunities available in the production of wheat. With the consistent increase in the population across the world, it has become more important to explore opportunities to increase the overall productivity of wheat yield. However, there are several challenges that come along with these opportunities.

To accommodate the demand of people, it becomes extremely imperative to significantly address these challenges effectively and efficiently. It is worth noticing that the challenges of the production of wheat depends from country to country, contingent on the local environment and other factors, in addition to the climate change impact. Therefore, taking into consideration all this will be helpful in successfully tackling these.

REFERENCES

1. (2012.) *The Adaptation Challenge*. [e-Book] Oxfam research reports. Available at: https://oi-files-d8-prod.s3.eu-west-2.amazonaws.com/s3fs-public/file_attachments/rr-adaptation-challenge-russian-federation-agriculture-112012-en_0.pdf (accessed on 7 July 2021).
2. Barma, N., Hossain, A., Hakim, M., Mottaleb, K., Alam, M., Reza, M., & Rohman, M., (2019). Progress and challenges of wheat production in the era of climate change: A Bangladesh perspective. *Wheat Production in Changing Environments*, 615–679.
3. Blumenthal, G., (2014). *Challenges and Opportunities for the U.S. Wheat Industry*. [online] Wheatfoundation.org. Available at: https://wheatfoundation.org/wp-content/uploads/2017/01/NWF-WPI-Wheat-Report-on-Behalf-of-Industry-Final-10072014.pdf (accessed on 7 July 2021).
4. Borlaug, N., (2021). *Wheat Breeding and its Impact on World Food Supply*. [e-Book] Third international wheat genetics symposium. Available at: http://herba.msu.ru/shipunov/school/univ_110/papers/borlaug1968_wheat_breeding.pdf (accessed on 7 July 2021).
5. Curtis, T., & Halford, N., (2014). Food security: The challenge of increasing wheat yield and the importance of not compromising food safety. *Annals of Applied Biology, 164*(3), 354–372.
6. Deng, X., Shan, L., Inanaga, S., & Ali, M., (2003). Highly efficient use of limited water in wheat production of semiarid area. *Progress in Natural Science, 13*(12), 881–888.
7. Duveiller, E., Singh, R., & Nicol, J., (2007). The Challenges of Maintaining Wheat Productivity: Pests, Diseases, and Potential Epidemics. *Euphytica, 157*(3), 417–430.
8. Ellis-Jones, J., Mekonnen, K., & Gebreselassi, S., (2013). *Challenges and Opportunities to the Intensification of Farming Systems in the Highlands of Ethiopia: Results of a Participatory Community Analysis*. [e-Book] Available at: https://cgspace.cgiar.org/bitstream/handle/10568/33789/ethiopia_pcareport_jul2013.pdf?sequence=10&isAllowed=y (accessed on 7 July 2021).
9. Hawkesford, M., Araus, J., Park, R., Calderini, D., Miralles, D., Shen, T., Zhang, J., & Parry, M., (2013). Prospects of doubling global wheat yields. *Food and Energy Security, 2*(1), 34–48.

10. National Association of Wheat Growers |. (2021). *Wheat Crop Challenges* | National Association of Wheat Growers. [online] Available at: https://www.wheatworld.org/wheat-101/research/wheat-crop-challenges/ (accessed on 7 July 2021).
11. Nature, (1910). *Problems of Wheat Growing, 85*(2141), 57.
12. Pingali, P., (2021). *Global Wheat Research in a Changing World: Challenges and Achievements*. [e-Book] Available at: https://ageconsearch.umn.edu/record/23726/files/fa99pi01.pdf (accessed on 7 July 2021).
13. Potter, B., (2021). *USDA Crop Progress: Corn Plantings Race Ahead*. [online] Farm progress. Available at: https://www.farmprogress.com/crop-report/usda-crop-progress-corn-plantings-race-ahead (accessed on 7 July 2021).
14. Ramadas, S., Kiran, K. T., & Pratap, S. G., (2020). Wheat production in India: Trends and prospects. *Recent Advances in Grain Crops Research*.
15. Rauf, S., Zaharieva, M., Warburton, M., ZHANG, P., Al-Sadi, A., Khalil, F., Kozak, M., & Tariq, S., (2015). Breaking wheat yield barriers requires integrated efforts in developing countries. *Journal of Integrative Agriculture, 14*(8), 1447–1474.
16. Rawson, H., Carberry, P., Hossain, A., & Saifuzzaman, M., (2011). *The Challenge to Increasing Wheat Production*. [online] www.researchgate.net. Available at: https://www.researchgate.net/publication/266080751_The_challenge_to_increasing_wheat_production (accessed on 7 July 2021).
17. Tembo, B., Sibiya, J., Tongoona, P., Melis, R., & Mukanga, M., (2021). *Farmers' Perceptions of Rain-Fed Wheat Production Constraints, Varietal Preferences and their Implication to Rainfed Wheat Breeding in Zambia*. [e-Book] Available at: https://arpgweb.com/pdf-files/jac2(12)131-139.pdf (accessed on 7 July 2021).

Chapter 6

Climate Change and Food Security with Emphasis on Wheat

CONTENTS

Environment comes as an important aspect for food security affecting farmers at various levels. This chapter briefly discusses the change in climate concerning agriculture and farming. In length, various mechanisms for farmers such as sustainable farming are being explained, creating numerous opportunities for farmers in terms of their financial stability and dealing with climate variability. Furthermore, solutions for adapting to the current environment at different levels such as state and national level are examined, and forestry systems in world view are addressed.

6.1. INTRODUCTION

On various timescales, human beings have learned to live with climate variability, from daily to decadal. Though, by a rise in global mean temperature, the climate variability we are accustomed to be changing quickly, accompanied due to increasing greenhouse-gas concentrations in the atmosphere. With the current climate variability, the poor in developing countries who already have difficulties in coping will be even more vulnerable. To the emissions of greenhouse gases, they are the ones who contribute the least, and to cope with changing climate with few financial or technical resources yet need to learn. In the light of climate change, this article first discusses the multiple aspects of food security. On crop production, the next part looks at impact at the differential spatial scales. For the food security of smallholders, the adaption to climate variability is most urgent, while climate prediction and longer-term climate-change-impact assessments constitute the basis for adaptation measures. In Morocco, this is discussed with an example of study and a focus on the use of climate prediction and information. In the agriculture sector, the article concludes with a discussion on adaptation and mitigation measures that are often mutually supportive.

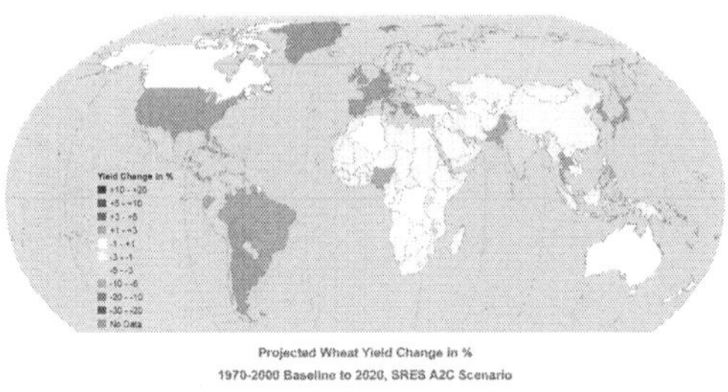

Figure 6.1. Climate change and food security with emphasis on wheat.

Source: Image by Wikimedia commons.

By impacting the basic human needs, climate change affects the livelihoods of poor and rich, including food, clothing, and shelter requirements. The four components of food security are food availability, food access, food utilization, and food production system stability, which are the heart of the mandate of the Food and Agriculture Organization of the United Nations (FAO).

By the climate, all four components are affected (FAO, 2008), but the food availability is most intimately associated with climate and its changes, from crops to animal products, marine and aquaculture products, and wood and nonwood products from forests.

If a system of food allocation, whether it is through market or not, is negatively affected, food access is impaired, and food security is compromised even when the production is sufficient. In many countries of the world, urbanization is rapidly taking place, creating a category of urban poor who do not themselves farm and are very vulnerable to climate change.

Due to climate change, projections of increased pests and diseases have an important implication for nutrition. Crops, livestock, fish, and humans will get affected by the new risks. For household members when human health is compromised, particularly that of women who prepare foods and the capacity to use food effectively is dramatically lowered.

With degraded hygiene in preparing food under limited freshwater availability or food-storage ability food safety may also be compromised due to warmer climate. Due to shrinking food biodiversity and excessive dependence on a few staple foods malnutrition may also increase.

On the food production system stability, the changes in climate variability have a direct implication. The frequency and intensity of extreme events have been increased, such as drought and flood would be a great threat to stability, whether the impact is domestic or through the global food market.

From complex interrelations between political conflicts and migration in a context of increased competition for limited resources, the frequency and magnitude of food emergencies might increase.

Under the climate change food availability and agricultural production are being discussed in the Chapter 5, "Food, Fibre and Forest Products" of the second volume of the Fourth Assessment Report of the Intergovernmental Panel on Climate Change (IPCC, 2007) and a number of other studies that have been published since then (e.g., Cline, 2007; Lobell et al., 2008).

Basically, in the cold areas, crop yields will increase where low temperature currently limits crop growth. On the other hand, in warm environments, heat stress on crop and water availability will lead to a decrease in yields. Globally, if night temperatures increase and averages rise by more than a few °C food production may increase, but a net negative impact is expected.

Moreover, on the global food production, there is a potential negative impact that there is pressure from the projected increase in population in most developing countries. From a recent FAO study which produced a typology of vulnerable countries to climate change, this is illustrated in a plot of net primary production of biomass, a biophysical indicator of potential agricultural production.

From the temperature, precipitation and population projections, net primary production per capita in 2030 was calculated. It is appeared in a way that only the parts of Europe, form purely biophysical, geophysical and demographical factors, the Russian Federation and Japan may benefit from increased productivity due to warming in the next couple of decades.

6.2. USE AGRICULTURE TO FIGHT CLIMATE CHANGE

In the history, 2019 will go down as the most difficult planting season for North American farmers, with over 10 million acres of crops going unplanted due to extreme weather conditions. Almost every month, at the same time, farmers in Punjab, in India, are experiencing rain showers and, for the first time in its history, more humid air is leading to greater pest infestations.

By the farmers, the effects of climate change can be felt daily, but to address this catastrophic threat, very few solutions have been discussed. Though, to reducing the amount of greenhouse gases trapped in the atmosphere: agriculture there is one widely unknown solution.

Into the crop production system reducing tillage, expanding crop rotations, planting cover crops, and reintegrating livestock have proven to reduce agriculture's own footprint as well as capture the excess carbon generated by other industries. After capturing the carbon then it is converted into the plant material and soil organic matter, improving soil health and increasing the ability to produce food on the land in the future.

As well as input cost has been reduced by these practices. Adopting these practices seems like the obvious choice, so you might be curious why a majority of farmers globally have continued using traditional agriculture practices.

Answer to that is in two different geographies farmers lifelong is reflect the difficult journeys that begin to adopt these management practices and the multiple barriers that are being faced a long way. With no guarantee on the immediate success pursuing a greener production system requires farmers to embark on uncharted territories.

Figure 6.2. Use agriculture to fight climate change.

Source: Image by the American society of agronomy.

During the transition process, farmers usually experience the decreased yields, as they gain the required experience to learn and perfect the implementation of more regenerative and beneficial practices.

A difficult challenge is to overcome which is being possessed by a decrease in the production, especially for the Indian farmers who already have a hard time competing with developed nations, where subsidies have artificially driven down the price of agriculture produce.

By keeping prices artificially low, the government's import and export policy decisions, which heavily favor consumers over producers, also have a large impact on the ability of farmers in India to adopt more sustainable practices.

Systematic financial challenges are being faced by the farmers in the US, like at a reasonable price difficulty accessing sustainable inputs. To their bottom-line farm input suppliers are highly concentrated, exerting significant pricing power and making systems innovation unattractive.

Along with the required upfront costs, increase the need for access to external capital, it will result in high operating cost. Though, with the most traditional capital, low-return production systems are most available to the farmers. In short, generally, the systems that are least regenerative, emit the most greenhouse gases, and result in the most land degradation, are the most likely to have access to capital.

In the availability of external capital subsidies and regulations play a role, since these cash streams serve as a de-risking mechanism for finance players. The US government policies focus on ensuring a stable food supply rather than on the nutritional value or environmental impacts of the food being produced, such as a federally subsidized crop insurance.

This focus is same in India that quantity over the quality, where decision-makers have rolled out an initiative to increase food production by the year 2050. This goal has become the cornerstone of India's national policy and the metric for measuring farmers' success.

In the lack of data, another key challenge is being faced and available information on these management practices as well as the lack of ways to measure which of these new, innovative systems store the most carbon in the earth's soil. For sequestering carbon, such a tool would open the door to private parties compensating farmers.

In the meantime, for new governments' regulations, along with growers' low profit margins, stifle farmer innovation, limiting the ability to test new practices for current farmers or creative thinkers to join the industry.

For example, in the US, subsidized crop insurance inflates the value of farmland and locks producers into a low risk, low reward system, making it hard for new smallholder farmers to enter the business or current farmers to walk away from the easy revenue.

In addition to it, the tax code makes it financially efficient to sell or exchange farms and farmlands only after death in the US. Along with this to shelter gains realized on non-agricultural real estate, further increasing the barrier to entry farmland investments may be used.

For farmers there is a lack of financial incentives to peruse innovation in both US and India, while for fundamental research public funding was there and to agricultural practices the application of research has been reduced in real terms.

Through the mono-cropping and chemical usage starved of funds, the public research system has taken the easier path and focused on identifying ways to maximize farm yields, both of which increase agriculture emissions. This research further dissuades farmers from adopting sustainable practices.

To focus on increasing the soil health one way to incentivize farmers through de-commoditizing production and making it easier to identify and track producers, as food makes its way from farm to table. For greener farming, doing so would unlock consumer demand as an incentive. For the food products, if consumers are willing to pay a higher price that have been grown in an environmentally friendly manner, the economics of the system could be turned on its head.

Unfortunately, to reward the farmers, there is no pathway left for consumers for establishing more favorable production systems, as US crop buyers currently blend and commoditize production. In India, where disposable income is fairly limited, increasing the price of foods could limit individuals' ability to purchase foods, amplifying the fear of food inflation, a problem the government dreads.

It has been seen that when trying to adopt sustainable management practices, it has been experienced that first-hand farmers face a variety of misaligned incentives and some of which require the whole ecosystem of actors to work together to overcome.

Each country has its own climate and set of challenges which is being possessed by the plot of land, and it is clear that there is no one size fits all solution. Though, to solve a portion of the climate change problem addressing these barriers is critical to unlocking agriculture's ability.

6.3. WAYS FARMERS CAN ADAPT TO CLIMATE CHANGE AND GENERATE INCOME

To the farmers around the world, climate change poses a real threat. On the good weather, including high and low temperatures, rainfall, wind intensity, and many other variables, agriculture is highly dependent. By 2050 and by 17%, estimates show that climate change might reduce global agriculture productivity. On the agricultural exports there are countries which are highly dependent like Brazil, this poses a real problem. In the rainfall patterns, research shows that climate change-driven changes could severely reduce the harvest of healthy soybeans, one of Brazil's biggest export commodities. By 95%, studies show that temperature increases could reduce the area suitable for producing coffee, an important export commodity, in three Brazilian states. Farmers can adapt the good news is that a new working paper published by WRI Brazil, GIZ, and the Brazilian Coalition on Climate, Forest, and Agriculture, with the contribution of several experts and scientists, presents evidence and proposes solutions for reducing climate change risks to farmlands. In Brazil, these include policies already in place that are still relatively unknown, but that could be expanded in Brazil, or replicated in other countries. To climate change at the same time, here is a selection of four sustainable ways farmers can produce more food and adapt.

6.3.1. Integrate Crop – Livestock – Forestry Systems

The greater its ability on average to adapt to climate change, the more diverse an agricultural system. Integrated systems combine them into one of four combinations: crops and livestock; crops and forestry; livestock and forestry; or crops, livestock and forestry, instead of focusing the farm on one kind of production (crops or livestock or forestry). The next question is that how does this work. According to the Brazilian Agricultural Research Corporation (Embrapa), the Brazilian Agricultural Research Corporation, integrated systems can produce food, energy, fibre, timber and non-timber forest products in the same area, at the same time or in rotation. Adaption benefits can be provided by the integrated systems. For every component analysed, research showed integrated systems that can make farms more resilient, and by reducing local temperature and increasing precipitation and water availability they can improve the local microclimate, reduce the impact of extreme weather events on crops, livestock and other products; reduce soil erosion; improve productivity; and provide additional socioeconomic benefits by increasing the number of products farmers can produce for subsistence or to sell.

Figure 6.3. Integrate crop-livestock – forestry systems.

Source: Image by Sentient media.

For example, with forestry in Brazil INOCAS is integrating pastures. To improve soil quality, the company plants native palm trees in pastures and create shade for the pasture's animals. Vegetable oil is being produced by the trees. To handle high temperatures, and provides new sources of income, this integrated system improves the land, makes livestock better able.

6.3.2. Rehabilitate Degraded Pastures

On Brazilian farms, pasture degradation is a major problem. To erosion degraded lands are prone and so retain less water, have less nutritious grass for feeding animals and contribute to low-productivity livestock production.

To rehabilitate degraded pastures, there are several different ways. For farmers, the most conventional way involves applying fertilizers, but this is not always sustainable, as pastures usually require new fertilizer applications every four or five years. By planting native forage or grass, or by introducing trees in the pasture to avoid soil erosion, degraded pasture can be more sustainably recovered.

To new rehabilitation of pasture can contribute, more sustainable ways of raising animals, as demonstrated by the work of Fazenda Ecological. With the trees, this combines grazing to improve the health of the soil and the wellbeing of the animals while mitigating carbon emissions.

Climate adaptation benefits are being provided rehabilitation of degraded pastures, including reduced local temperatures, increased air humidity, better resistance against heatwaves and drought, and more resilience against natural disasters. On soil erosion and water availability it also has a positive effect.

6.3.3. Plant Agroforestry Systems

In an intentionally designed system agroforestry integrates trees and crops. For a particular purpose, every plant is selected in an agroforestry system, species are selected so that plants will not compete but collaborate. To productive all year long, this diversity of crops and trees allows all year long, so that small farmers can earn income in all seasons.

From agroforestry cocoa is one commodity get benefits. Under the shade of other trees, coca plants can grow better and be even more resilient. To support 20,000 farmers who produce organic and/or high-quality cocoa in agroforestry systems on 58,000 hectares, in Peru, a public-private partnership seeks.

For climate change adaptation in agriculture, agroforestry systems are an important tool. For the local climate, the working paper finds that agroforestry produces adaptation benefits, including reducing the impact of five types of extreme weather events evaluated by the study (drought, heatwaves, cold waves, heavy rain, and floods), improving soil and water availability, attracting pollinators and improving biodiversity.

6.3.4. Pursue Sustainable Forestry

Environmental benefits are being offered by the trees planted sustainably, such as capturing greenhouse gases and protecting the soil, as well as the potential for economic gain through the commercialization of timber and nontimber forest products. In 2018, the WRI report is being highlighted 14 companies that are investing in the business of planting trees.

In almost every factor aside from being a good investment, sustainable forestry provides positive adaptation effects, with an important exception, as the risk of forest fires could be increased. Densely planted trees and flammable species such as eucalyptus can cause fires to quickly spread.

For Brazilian farmers, restoration and reforestation can be important tools. With the natural vegetation, the Brazilian Forest Code states that a portion of all rural properties must be covered. Estimates indicate that 21 million hectares of Brazilian forest (51 million acres) are degraded or deforested and need to be restored. To provide economic and environmental benefits where plantations are not allowed, but an important portion could be planted with sustainable forests. For timber planting, native species or other purposes could become an important source of income for farmers while helping rural lands and production adapt to climate change.

6.3.4.1. Lessons for Brazil and the World

For farmers practices like those described above provide a possible path around the world move to more resilient, low-carbon agriculture, so they can more sustainably produce enough food to feed the world while restoring the land. By finding a more resilient and sustainable model for agribusiness, it is crucial to protect farms and farmers, in a world where the impacts of climate change are already arriving.

In two important Brazilian policies, these examples of sustainable farming and land management practices are already included. The Low Carbon Agriculture Plan, known as Plano ABC, and the National Plan for Recovery Native Vegetation, known as Planaveg. For the agricultural models proposed by these plans should be expanded in Brazil the climate and economic benefits outlined.

For the climate change adapting farms is good policy not only for the environment but also the economy, financial institutions, and insurance companies. For the financial institutions reducing risk from extreme weather events is crucial dealing with agribusiness credit and insurance companies, reducing risk losses and improving loan repayment.

For the farmers models like agroforestry and sustainable forestry plantations show a possible path around the world move to low-carbon agriculture and a more resilient agriculture, producing the necessary food to feed the world while restoring the land. By finding a more resilient and sustainable model for agribusiness, it is crucial to protect farms and farmers. In a world where the impacts of climate change are not just imminent but arriving.

6.4. AGRICULTURE'S ROLE IN ADDRESSING A CHANGING CLIMATE

For the humanity, climate change is one of the most pressing challenges, and agriculture feels its effects in profound ways.

By extreme weather conditions, farmers are particularly impacted, which include drought, severe heat, flooding, and other shifting climatic trends. For farmers, these all pose challenges, as they work to grow enough food, which is why we're devoted to finding ways to transform agriculture to be part of the solution in addressing climate change. By 2030, we commit to achieve a 30% reduction of the field greenhouse gas footprint of our farming customers.

6.5. FROM CONTRIBUTOR TO MITIGATOR

To the climate change, agriculture is a contributor, in curbing greenhouse gas (GHG) emissions like carbon dioxide, methane, and nitrogen oxide the industry plays a role that contribute to climate change.

Farmers are taking steps toward a carbon-zero future, to help ensure a more sustainable future and using cutting-edge tools and farming practices to remove as much, if not more greenhouse gases from the atmosphere than a farmer emits.

Figure 6.4. Agriculture's role in addressing a changing climate.

Source: Image by Pixabay.

Including digital farming, the development of climate-smart solutions, and the improved plant breeding technologies will help reduce agriculture's impact on climate change in the future, and here are some of the many existing practices that are making a positive impact today.

6.5.1. Reducing Tillage

Farmers reduce the need to till, with the better weed control solutions, as decreasing tractor passes over the field and allowing for less soil disruption. It is better able to store carbon when the soil is left untilled, as well as nutrients and water, and this only curb greenhouse gas emissions and fossil fuel use.

6.5.2. Increasing Efficiency

To have a more intimate and informed understanding of what's happening in their field's digital tools and precision agriculture techniques enable the farmers. For example: Bayer is working on software platforms that offer monitoring tools that help farmers use pesticides more efficiently, reducing greenhouse gas emissions and pesticide runoff into water.

To apply pesticides precipitation alerts let farmers know which fields may be too wet or too windy so they can avoid costly treatments and unnecessary fossil fuel use. Real time health images are being provided by the satellite and drones that enable farmers to identify areas of crop stress or pest infestations so that corrective actions can be taken quickly and efficiently. To grow enough food on less acreage, these tools help the farmers to optimize the land use, offering the potential to reduce the number of acres needed to feed a growing population.

6.5.3. Optimizing Nutrients

To enrich crops by helping plants harness and use nutrients and water more efficiently, agriculture is also pioneering the use of microbes. to create microbes that will help crops like corn, wheat, and rice extract nitrogen from the air to use on their own Joyn Bio, a joint venture between Bayer and Ginkgo Bio works, is working on this. for man-made fertilizer this would radically reduce the need and for agriculture a powerful step towards a carbon-zero future.

6.5.4. Halting Deforestation

When agricultural expansion leads to deforestation or draining of wetlands, the largest source of atmospheric carbon related to farming occurs, which reduces the ability of the natural ecosystem to absorb and store carbon.

On the less land solutions are being developed that will help the farmers to grow enough and agriculture is working to preserve natural habitat, even finding ways to help farmers identify areas of their land that would add more value to their operation as a biodiversity sanctuary to support animals, pollinators, and their surrounding environment.

6.5.5. Helping Farmers Manage Changes in Weather

On the climatic conditions' smallholder farmers are particularly dependent to grow their crops. Extreme weather conditions could threaten their entire harvest, with an increasingly unpredictable climate. To empower smallholders with access to tools, insights, and resources to mitigate these effects, Bayer is committed to finding innovative ways so they can cultivate their crops in spite of climatic challenges.

Figure 6.5. Helping farmers manage changes in weather.

Source: Image by Flickr.

6.5.6. Withstanding Floods

To thrive all crops, need water, but too much water can be devasting. For over 3.5 billion people, rice is an important staple food, who rely on approximately 200 million small farmers to grow it.

In Vietnam, flooding throughout the Mekong Delta has been a considerable challenge for smallholder farmers to cultivate their crops while meeting growing demand. The sea can flood the rice paddies and salinize (or add salt to) the soil, if the rains are late.

Bayer's "Much More Rice" program offers a comprehensive solutions program featuring innovative technologies such as hybrid seed that can withstand up to 14 days of being submerged, in order to help smallholders, withstand and overcome these challenges.

On the other side of the spectrum, it has been seen that the climate change can also create conditions that lead to drought. In the Midwestern United States, wheat farmers require about 50 centimetres (almost 20 inches) of precipitation per year, but in some years, they only get half of that. Recent studies estimate that wheat harvests could decline by 6% with every degree of climate warming, with the drought as a considerable factor.

60% – The demand for wheat will increase by approximately 60% by 2050.

To maintain moisture in the soil, many farmers use the reduced tillage methods, but to compete with wheat plants for light, nutrients this can create opportunities for invasive weeds and the very moisture farmers are working to conserve. There is the development of new and more effective crop protection solutions that are helping farmers, who are managing their crops without leaving them vulnerable to additional threats such as weeds, insects, or disease.

6.5.7. One Challenge with Countless Solutions

Climate change is an opportunity for people to reimagine which people can accomplish through the agriculture, as climate change is a major challenge. Moreover, to reduce agriculture's impact on the environment, there is development of new solutions, as how to shape agriculture to become part of the solution is also being explored. In spite of many challenges, work is done to accomplish to empower farmers with the tools they need to grow their crops toward addressing climate change.

6.6. TOWARDS SUSTAINABLE, PRODUCTIVE AND CLIMATE-FRIENDLY AGRICULTURAL SYSTEMS

For the agriculture sector, climate change has created challenges that will continue to do so. In temperatures, climate change-induced increases, and rainfall variation and the frequency and intensity of extreme weather events are adding to pressures on global agricultural and food systems.

In most of the regions, climate change is expected to negatively affect both crop and livestock production systems, although from the changing conditions, some countries may actually get benefit. To resource problems, the changing climate is also adding, such as water scarcity, pollution, and soil degradation.

The significant share of the greenhouse gas (GHG) emission is being contributed by the agriculture that are causing climate change – 17% directly through agricultural activities and an additional 7–14% through changes in land use. It is considered as an important part of the solution.

Nitrous oxide emissions from soils, fertilizers, manure, and urine from gazing animals are the main direct agricultural GHG emissions and methane production by ruminant animals and from paddy rice cultivation. Then the carbon dioxide both of these gases have a significantly higher global warming potential.

Figure 6.6. Towards sustainable, productive, and climate-friendly agricultural systems.

Source: Image by Pxhere.

To adapt to climate change self-initiated by the farmers while decreasing the GHG footprint of agriculture are unlikely to be sufficient, given uncertainties surrounding the timing and nature of climate change. In the global emission, the share of agriculture, and to underline the necessity of stronger, more effective policies, the need for further global mitigation efforts, and continued projected agricultural emissions growth in many countries all combine.

What Should Policy Makers Do?

Within and beyond the agriculture sector policy reforms are needed to strengthen farmer incentives to achieve sustainable productivity growth without sacrificing climate change mitigation and adaptation.

6.6.1. At International Level

At the 2015 United Nations climate change conference – COP implement the Paris agreement. For the emissions reduction both the text and the country level strategies are outlined in the form of Intended Nationally Determined Contributions (INDCs), as to sustainable food production recognize the threat which climate change poses, and to climate change offer valuable opportunities for agriculture and the food chain to be an active part of the solution.

6.6.2. At National Level

There are wider social, economic and environmental policy settings such as trade, investment, infrastructure, and education policies which should consistently support sustainable productivity growth, in combination with adaptation and mitigation efforts. By the farmers inconsistent or contradictory signals, such as import restrictions that protect water-intensive crops, can exacerbate poor choices. It has been said that the marginal fine-tuning of existing agricultural policies improving overall policy coherence will be more effective.

6.6.3. At Sector Level

To encourage the unsustainable intensification, there is a need to reform misaligned and distortive agricultural policies and along with this, the overuse of natural resources and potentially damaging inputs. Across the OECD are more than half of the agricultural support continues to be potentially damaging to the environment, while measures targeting sustainable productivity or climate change goals remain marginal. With a view to their eventual elimination oversubsidized insurance, market price support and input subsidies should be reduced.

In the research and development (R&D), further investment is needed to spur innovation that can improve sustainable productivity growth. Private sector innovation may be facilitated further by the governments, for example, addressing investment barriers that impede R&D, ensuring that private knowledge is disseminated, and encouraging – where appropriate – public-private partnerships for R&D that have public goods outcomes.

To address the climate change policies that are aiming should emphasize outcome-based farmer incentives and to achieve sustainable productivity growth through mitigating and adaptive practices knowledge transfer systems that enhance farmer capacity.

As opposed to practices sustainability performance should be targeted by the financial incentives where possible. to adopt measures that have high upfront costs, untargeted incentives can encourage farmers, or that are socially beneficial but costly at a private level.

On resource use efficiency, governments should ensure the provision and dissemination of relevant and up-to-date information and in adaptation and mitigation measures risk management to help farmers and other private agents make informed investments. Strengthening access to knowledge and risk management mechanisms is key to increasing the adoption of sustainable and productive practices. To streamline adaptation and mitigation advisory, it is often advisable that actions into existing institutions and to coordinate such actions with the private sector.

6.7. FOOD SECURITY

The great majority of the world population get feeds by the current food system and supports the livelihoods of over 1 billion people. By greater use of nitrogen fertilizers (increase of about 800%) and water resources for irrigation food supply per capita has increased more than 30%, since 1961.

Though, there is an estimation that 821 million people are currently undernourished, 151 million children under five are stunted, 613 million women and girls aged 15 to 49 suffer from iron deficiency, and 2 billion adults are overweight or obese. From the nonclimate stressors, the food system is under pressure and also from the climate change. The four pillars of the food security get impacted by these climate and non-climate stresses.

Figure 6.7. Food security plays an important role in agriculture.

Source: Image by Snappygoat.com.

Food security is already get affected by observing the climate change through increasing temperatures, changing precipitation patterns, and greater frequency of some extreme events (high confidence). From the other factors affecting crop yields it has been seen that studies that separate out climate change have shown that yields of some crops (e.g., maize and wheat) in many lower-latitude regions have been affected negatively by observed climate changes, while in many higher-latitude regions, yields of some crops (e.g., maize, wheat, and sugar beets) have been affected positively over recent decades.

By drying, warming has been compounded that has caused the large negative effects on yields in parts of the Mediterranean. Climate change is affecting food security in drylands, based on the indigenous and local knowledge (ILK), particularly those in Africa, and high mountain regions of Asia and South America.

By the projected future climate change, food security will be increasingly affected. Due to the climate change the across Shared Socioeconomic Pathways (SSPs) 1, 2, and 3, global crop and economic models projected a 1–29% cereal price increase in 2050, which would impact consumers globally through higher food prices; regional effects will vary (high confidence).

At-risk particularly there are low-income consumers, with the models projecting increases of 1–183 million additional people at risk of hunger across the SSPs compared to a no climate change scenario (high confidence).

For the crop productivity, the increased CO_2 is projected to be beneficial at lower temperature increases, it is projected to lower nutritional quality (high confidence) (e.g., wheat grown at 546–586 ppm CO_2 has 5.9–12.7% less protein, 3.7–6.5% less zinc, and 5.2–6.5% less iron). In many regions, distributions of pests and diseases will change, affecting production negatively. The risks of the food system disruptions are growing, as it is given increasing extreme events and interconnectedness.

To the climate change vulnerability of the pastoral systems is very high. In more than 75% of the country's pastoralism is practiced between 200 and 500 million people, including nomadic communities, transhumant herders, and agropastoral. In the Africa impacts in pastoral system include lower pasture and animal productivity, damaged reproductive function, and biodiversity loss. By the nonclimate factors pastoral system vulnerability is exacerbated.

A key component of the healthy diet is fruit and the vegetable production which is also vulnerable to climate change (medium evidence, high agreement). Under the high temperatures there is declines in the yields and crop sustainability are being projected, especially in tropical and semitropical regions.

Fruit set has been reduced by the heat, and it also speeds up development of annual vegetables, which will result in yield losses, impaired product quality, and increasing food loss and waste. To the greater annual yields, longer growing seasons enable a greater number of plantings to be cultivated and which also contribute. Though, to produce a viable harvest some fruits and vegetables need a period of cold accumulation, and warmer winters may constitute a risk.

Gender and equity dimensions are the thing which the food security and climate change have. In food security, women play a key role worldwide, also the regional differences exist. Among the diverse social groups, it gets impacted by the climate change depending on the age, ethnicity, gender, wealth, and class.

On the livelihoods of poor and vulnerable community's climate extremes have immediate and long-term impacts on livelihoods, which is contributing to greater risks of food insecurity that can be a stress multiplier for internal and external migration. For the decision making empowering women and rights-based approaches can create synergies among household food security, adaptation, and mitigation.

To the advance adaptation, many practices can be optimized and scaled up throughout the food system (high confidence). For the tolerance to heat and drought supply-side options include increased soil organic matter and erosion control, improved cropland, livestock, grazing land management, and genetic improvements.

To reduce it has been said that diversification in the food system which is a key strategy. There is demand-side adaption, such as adoption of healthy and sustainable diets, in conjunction with reduction in food loss and waste, can contribute to adaptation through reduction in additional land area needed for food production and associated food system vulnerabilities. To enhance the food system resilience ILK can contribute.

To the food system about 21–37% of total greenhouse gas (GHG) emissions are attributable. Generally, these are from agriculture and land use, storage, transport, packaging, processing, retail, and consumption. Within the farm gate this estimate includes emissions of 9–14% from crop and livestock activities and 5–14% from land use and land-use change including deforestation and peatland degradation (high confidence); 5–10% is from supply chain activities (medium confidence).

From the food loss and waste, this estimate includes GHG emissions. During the period 2007–2016 within the food system, the major sources of emissions from the supply side were agricultural production, with crop and livestock activities within the farm gate generating, respectively. It has been seen that CO_2 emissions linked to relevant land-use change dynamics such as deforestation and peatland degradation.

From the IPCC AR5, by using 100-year GWP values this implies that total GHG emissions from agriculture were 6.2 ± 1.4 $GtCO_2$-eq yr^{-1}, increasing to 11.1 ± 2.9 $GtCO_2$-eq $yr.^{-1}$ including relevant land use. Due to increasing demand based on population and income growth and dietary change, these are likely to increase by about 30–40% by 2050 without intervention.

By reducing crop and livestock emissions, supply-side practices can contribute to climate change mitigation, sequestering carbon in soils and biomass, and by decreasing emissions intensity within sustainable production systems (high confidence). From crop and livestock activities total technical mitigation potential agroforestry is estimated as 2.3–9.6 $GtCO_2$-eq $yr.^{-1}$ by 2050.

In the cropping system options with large potential for GHG mitigation include soil carbon sequestration (at decreasing rates over time), reductions in N_2O emissions from fertilizers, reductions in CH_4 emissions from paddy rice, and bridging of yield gaps.

For mitigation in livestock systems, options with large potential include better grazing land management, with increased net primary production and soil carbon stocks, improved manure management, and higher-quality feed. From the livestock, the reduction in GHG emissions intensity can support reductions in absolute emissions, provided appropriate governance to limit total production is implemented at the same time.

For reducing GHG emissions consumption of healthy and sustainable diets presents major opportunities from the food systems and improving health outcomes. There are some of the examples of healthy and sustainable diets are high in coarse grains, pulses, fruits and vegetables, and nuts and seeds; low in energy-intensive animal-sourced and discretionary foods (such as sugary beverages); and with a carbohydrate threshold. In the emissions this estimate includes from livestock and soil carbon sequestration on spared land, but co-benefits with health are not taken into account.

It has been said that mitigation potential of dietary change may be higher, but at broad scales, achievement of this potential depends on consumer choices and dietary preferences that are guided by social, cultural, environmental, and traditional factors, as well as income growth. In the transition, meat analogs such as imitation meat (from plant products), cultured meat, and insects may help to more healthy and sustainable diets, although their carbon footprints and acceptability are uncertain.

GHG emissions could get lower by the reduction of food loss and waste and improve food security. It has been seen that of the total food produced, the combined food loss and waste amount to 25–30%. Global food loss and waste equaled 8–10% of total anthropogenic GHG emissions (medium confidence); and cost about 1 trillion USD2012 per year (low confidence) during the year 2010–2016.

For the reduction of food loss and waste technical options include improved harvesting techniques, on-farm storage, infrastructure, and packaging. In the developed and developing countries, the causes of food loss and waste differ substantially, as well as across regions. To the global climate change responses, agriculture and the food system are key for it. There are the combination of the supply side actions such as efficient production, transport, and processing with demand-side interventions such as modification of food choices, and reduction of food loss and waste, reduces GHG emissions and enhances food system resilience (high confidence). The implementation of large-scale land-based adaptation is get combined by the such measures and can enable also and along with this mitigation strategies without threatening food security from increased competition for land for food production and higher food prices.

In the farm management without the combined food system which measures supply chains, and demand, adverse effects would include increased numbers of malnourished people and impacts on smallholder farmers. To address these effects just transitions are needed. Throughout the food system for adaption and mitigation, it enables the conditions need to be created through policies, markets, institutions, and governance (high confidence). Through the risk-sharing there is resilience to increasing extreme events can be accomplished and also the transfer mechanisms such as insurance markets and index-based weather insurance. The demand can be changed potentially by the procurement, health insurance incentives, and awareness-raising campaigns and contribute to lower GHG emissions. In the broader climate change policies without inclusion of comprehensive food system responses, the mitigation and adaptation potentials assessed in this chapter will not be realized, and food security will be jeopardized.

6.8. CLIMATE CHANGE THREATENS FOOD SECURITY

Climate change is the threat multiplier for hungry and undernourished people. The number of extreme weather-related disasters has doubled since the early 1990 s. An increase in food prices and a decrease in income has reduced the yields of major crops.

Low-income people have been harmed by these disasters disproportionately and also access to the food, and because of this, it has been chosen to focus on the connection between climate change and food security in the 2019 Global Hunger Index (GHI), prepared by Concern Worldwide and Welthungerhilfe. Any time soon, sadly, it has been seen that this is not a trend that appears to be going away.

In most land and ocean regions, climate models predict higher average temperatures by looking ahead, and in some areas, both heavy precipitation and an increasing probability of drought. For reducing the hunger, these are all additional challenges. In absolute terms, there is a reduction in the global hunger on a proportionate level since 2000, the number of people going to bed hungry is on the rise.

In reducing the hunger, the world has made gradual progress on a global scale since 2000, and this progress has been uneven.

The rate of undernourishment, child stunting, child wasting, and child mortality is been taken into account by the GHI to highlight where action is most needed to fight hunger. With 2019 the data is now available to have a look at some ways of the climate change threatens food security and how that may impact all.

6.8.1. Climate Change and Food Security: Fast Facts

1. Climate change is a threat multiplier for hungry and undernourished people.
2. Countries with high levels of hunger are often also highly vulnerable to climate change, and have a low capacity to adapt.
3. Climate change affects food production and availability, access, quality, utilization, and stability of food systems. In short, it impacts all aspects of the food system.

4. Extreme weather-related disasters are increasing and reduce the yields of major crops.
5. Higher levels of carbon dioxide reduce the nutritional value of crops.
6. The global food system contributes about one-third of greenhouse gas emissions. About ⅔ of food are lost and wasted from farm to table. These losses therefore exacerbate climate change without improving food security or nutrition.
7. Climate change and conflict combined destroy livelihoods, drive displacement, widen inequalities, and undermine sustainable development.
8. Ending hunger and undernutrition in a changing climate demands large-scale action.

Figure 6.8. Climate change threatens the food security.

Source: Image by Flickr.

6.9. CLIMATE CHANGE THREATENS AGRICULTURE AND FOOD PRODUCTION

In the atmosphere, higher temperatures, water scarcity, extreme events like droughts and floods, and greater concentrations of carbon dioxide have already begun to impact staple crops around the world. In the recent years, maize and wheat production has been declined due to extreme weather events, plant diseases, and an overall increase in water scarcity.

The unpredictable yield for cereal crops in semiarid regions of the world (like the Sahel region of Africa) is at least 80% the result of climate variability, according to the Food and Agriculture Organization of the United Nations.

Figure 6.9. Climate change threatens the agriculture and food production.

Source: Image by Imaggeo.

To the food security, rising sea levels pose a different threat, in other areas like Bangladesh and Vietnam. To flood coastal farmlands, the saltwater is more likely, and it is also more likely to kill off rice crops. On the Mekong Delta region, half of Vietnam's national rice production is centred which means that a flood in this region (roughly the size of Maryland) could impact both the country's food security and economy.

As, it has been seen that the causes of these issues are fairly consistent and the solutions are not. This means, in areas like the Horn of Africa and South-east Asia with the effects of climate change already apparent and there is a need to find the unique methods to mitigate disasters when they strike, and develop bespoke ways to reduce the impacts of hazards on the lives and livelihoods of these areas in the future.

6.9.1. Climate Change Limits Food Access

It has been said that if the food production may get affected by the climate change, then it stands to reason that it also affects food access. This simple supply-and-demand and demand has big impacts. For the food that is available, climate change and weather disasters can lead to inflated prices. With the urban poor spending up to 75% of their total budget on food alone, these price spikes leave the poorest households which are said to be most vulnerable.

Figure 6.10. Climate change limits the food access.

Source: Image by unsplash.com.

As it has been seen that the food system is increasingly dependent on one another, this means that more frequent and more extreme events in one region could disrupt clusters of food systems and even the global food system as a whole. Though, the areas least likely to adjust to a sudden event or shock which continue to be the ones disproportionately affected.

6.9.2. Climate Change Decreases Nutrition and Nutritional Value

The next concern becomes nutrition in many food-insecure areas. Until the next harvest families reducing their food consumption (often skipping one or more meals each day). With the climate change, there is reducing harvest, if there are fewer supplies, this means that the lean period may be extended or if it takes longer to get an adequate harvest. Due to the climate change, an estimated additional 175 million people could have zinc deficiencies, and an additional 122 million people could be protein deficient by 2050. The nutritional value of the food that is grown can be adversely affected by the climate change alternatively. The protein zinc and the iron content of the crops is being reduced by the higher carbon dioxide concentrations, as it is being shown in the studies. By 2050 among the other things, an estimated additional 175 million people could have zinc deficiencies and an additional 122 million people could be protein deficient. For their nutrition will communities relying largely on the plant harvest and feel this most acutely. On the livestock, this also has a ripple effect beyond the plant-based nutrition, which rely on the same resources as humans to eat, grow, and produce meat and/or milk. By the drought, livestock are also severely threatened, which is accounting for 36% of drought-related losses. Climate extremes also threaten fish populations, especially in areas like Southeast Asia.

Figure 6.11. Climate change decreases nutrition and nutritional value.

Source: Image by unsplash.com.

6.9.3. Climate Change Increases Food Waste

Either rain does not guarantee healthy crops. The toxic mild on the crops is being produced by the higher rainfalls or flooding. In the high drought areas, crops are grown that are then moved into the humid storage facilities, which are vulnerable to fungal infections or pests. The more climate changes and the more that extreme climate events become commonplace, the more food we lose on an annual basis.

Roughly one-third of the food produced by farmers is lost between the field and the market in low- and middle-income countries, according to the UN's Food and Agriculture Organization. Currently, the food system contributes 21–37% of greenhouse gases, meaning that these food losses add to the climate crisis but do nothing for food security or malnutrition levels.

6.10. CONCLUSION

Overall, climate change has a diverse effect on agriculture as a whole. Sustainable farming is the key to the future, resulting in increased production and benefiting farmers as well as decreasing food waste. Climate change can be treated effectively with the changes at state and national levels for agriculture by adopting different measures. Diverse adaptions are required to increase food security as well as to maintain forestry systems is of utmost importance in order to support farmers and for saving agriculture.

REFERENCES

1. Bayer Crop Science, (n.d). *Climate Change.* [online] Available at: https://www.cropscience.bayer.com/people-planet/climate-change (accessed on 7 July 2021).
2. Calmon, M., & Feltran-Barbieri, R., (2019). *4 Ways Farmers Can Adapt to Climate Change and Generate Income.* [online] World Resources Institute. Available at: https://www.wri.org/insights/4-ways-farmers-can-adapt-climate-change-and-generate-income (accessed on 7 July 2021).
3. Chakrabarty, M., (n.d). *Climate Change and Food Security in India* | ORF. [online] ORF. Available at: https://www.orfonline.org/research/climate-change-and-food-security-in-india/ (accessed on 7 July 2021).
4. Concern Worldwide, (2019). *How Climate Change Threatens Food Security (and why we're all at risk).* [online] Available at: https://www.concernusa.org/story/climate-change-food-security/#:~:text=Climate%20change%20affects%20food%20production,the%20nutritional%20value%20of%20crops (accessed on 7 July 2021).
5. Ipcc.ch. (n.d). *Chapter 5: Food Security — Special Report on Climate Change and Land.* [online] Available at: https://www.ipcc.ch/srccl/chapter/chapter-5/ (accessed on 7 July 2021).
6. World Economic Forum, (2019). *Here's how we Can Use Agriculture to Fight Climate Change.* [online] Available at: https://www.weforum.org/agenda/2019/09/here-s-how-we-can-use-agriculture-to-fight-climate-change/ (accessed on 7 July 2021).
7. World Meteorological Organization, (2009). *Food Security Under a Changing Climate.* [online] Available at: https://public.wmo.int/en/bulletin/food-security-under-changing-climate (accessed on 7 July 2021).

Chapter 7

Breeding Strategies and Quality Enhancement Under Climate Change

CONTENTS

In the chapter, breeding strategies and quality enhancement under climate change, the importance of breeding strategies in improving productivity yield is discussed. It also explains the role of crop adaptation to overall extreme climate stresses.

The chapter also highlighted the importance of exploring genetic resources to increase adaptation of wheat to climate change. It also shed some light on the wheat genetic resources evolution and their role in breeding. In addition, it also discussed some breeding methods and approaches to increase genetic gains. Finally, at the end, it explains several approaches to combat climate changes.

7.1. INTRODUCTION

Climate change can be termed as the changes in the statistical distribution of weather over a period of time that varies from decades to millions of years. It can be referred to a transition to in the average weather or a transition in the delivery of weather events around an average. The change in climate may be confined to a specific region, or might be the case across the world, and this kind of climate change has been recognizable (Sahney et al., 2010).

It is well known that since very long, farmers have adapting to the impacts of climate variability, but forecast of climate change pose a serious threat that will test farmer's potential to adapt and enhance their livelihoods (Adger et al., 2007). Climate change is a danger to food security and agriculture, and there is a pressing need to look for urgencies for future research.

The connection between agriculture, climate change, and food security, though, is a complex one that is also shaped by political decisions and economic policies. Adequate climate change research, thus, comprises researchers from a comprehensive spectrum of disciplines along with other shareholders.

It is important to note that maize and wheat are two of the most crucial cereal crops in the world and there is a growing concern about the impact of projected climate change on the productivity and production of these key cereal crops.

Climate change poses greater challenges to food security and the livelihood security of millions. Therefore, development and conveyance of enriched germplasm and risk-reducing management choices have the capability to balance some of the yield losses associated with climate change.

Communities may adapt in distinctive ways, embracing switching to water-efficient or heat and drought-tolerant crops better suited to a drier and warmer climate (Lobell et al., 2008) and/or diversifying livelihood strategies across livestock and crops (Seo, 2010).

Food security in a period of climate change may be probable if farmers transmute agricultural systems via the use of fertilizer and improved seed, in addition to better governance (Brown and Funk, 2008). Models of the global food economy propose that trade will also exemplify an imperative but not whole buffer against climate change persuaded yield effects.

The International Wheat and Maize Improvement Center (CIMMYT), along with national and international agricultural research institutes are actively working to advance wheat technologies for countries that are vulnerable to climate.

Wheat yields deteriorates at supra-optimal temperatures (Wardlaw et al., 1989; Reynolds et al., 1994) and substantial breeding effort will be needed to preserve productivity in regions that are nearer to the equator. However, wheat is comparatively well adapted to water shortages and is grown extensively in semiarid regions such as Australia, Central Asia, and throughout the Mediterranean region.

In regions that become gradually more arid, wheat may become more competitive in comparison to crops, such as maize, that are presently grown. Wheat breeding has had a substantial impact in marginal environments as well as temperate zones.

Some of the current effort has focused on breeding for earlier maturing cultivars that emit terminal heat stress and incorporate endurance to diseases related with warm humid environments (Joshi et al., 2007) as well as the extremely virulent Ug99 stem rust strain.

One of the best research approaches for wheat has been, and will continue to be, to alter the phenological pattern of the crop so that imperative growth stages do not correspond with worrying conditions or merely to finish the life cycle early before arise of stern stress conditions.

Another is to curtail the occurrence of stress with the progress of a good root system that, in an instance of drought, allows water to be retrieved deeper in the soil and, in the case of heat, permits transpiration rates that better tie with evaporative demand, thus supporting maximal carbon fixation with the extra advantage of cooler plants (Reynolds et al., 2010).

As it is well known that there is a time lag between technology development, on-farm adoption, and deployment of new varieties, current research also requires to emphasizes on policy options and institutional innovations that assist farmers' access to new as well as existing germplasm to improve local adaptive capacity to climate change.

7.2. CROP ADAPTATION TO OVERALL EXTREME CLIMATE STRESSES

With the gradual surge in the temperature of Earth, the climate suffers severe transition and becomes abiotically stressful. It is important to note that the environmental changes are extremely damaging and pose greater threats to naturally prevalent crop species. Under field circumstances, heat and drought are the most prevalent stresses and have a noteworthy influence on plants. It is stated that plants need an optimum temperature for their standard growth and blooming. Plant physiology is greatly induced by temperature variations. As heat stress impacts the overall production of grain production and yield, cold stress results in sterility, while drought stress negatively impacts the morpho-physiology of plants. These climatic problems extensively distress plant development and yield, generate giant responses, comprising biochemical, molecular, morphological, and physiological modifications. Overall, climate change and global warming both have some positive as well as negative effects on agricultural crops as well as on humans, as explained.

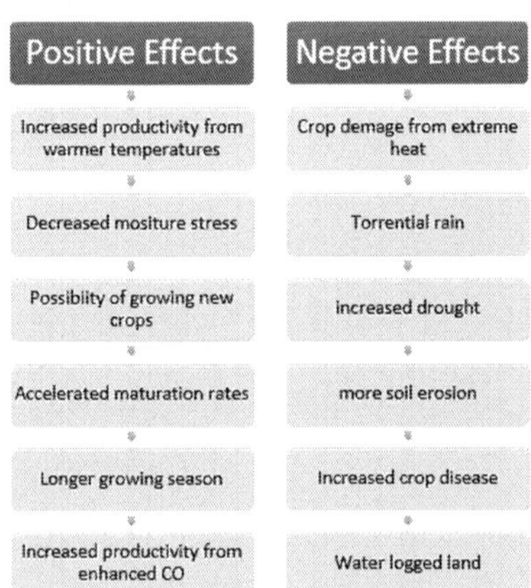

Figure 7.1. Overall positive and negative effects of climate change and global warming on crops and humans.

Source: www.ncbi.nlm.nih.gov.

In this manner, understanding the stress-resistance processes in plants has arisen as a very challenging task for plant scientists in a way to cultivate stress-resistant plants. The chief cereal crops across the world, such as rice, maize, and wheat, are critical to accommodate the daily demand of food. Out of them, wheat was the leading staple crop that has been cultivated on a large scale.

Wheat is harvested on 38.8% of total agricultural land worldwide and provides a significantly high concentration of proteins: 15% per gram in comparison to rice or maize, which only supplies 2 to 3%. Irrespective of large growing land globally, its productivity has been mainly less than the rice and maize. Reasonable reduction was predicted in wheat productivity with a 2°C surge in temperature.

Related research on environmental variability expected a 6% diminution in wheat yield. Challinor and his colleagues labeled that because of surge in temperature, the grain filling phase reduction is the primary cause of crop productivity cutback in changing climatic conditions. Thus, sustaining crop yield is an imperative task in current agriculture, and to produce stress-tolerant crop plants.

7.3. EXPLORING GENETIC RESOURCES TO INCREASE ADAPTATION OF WHEAT TO CLIMATE CHANGE

Figure 7.2. Adaption of wheat to changed climate.

Source: Image by Eos.org.

The CIMMYT in association with International Wheat Improvement Network (IWIN) partners with hundreds of wheat breeders across the world to offer new genotypes (~1,000 annually) to national programs as international public goods, through the following mechanisms (Braun et al., 2010):

1. Free exchange of germplasm with all national private and public breeding programs across the world, including agreements from genetic resource collections.
2. Centralized breeding hubs that emphasize on generic needs – i.e., yield stability, yield potential, genetic resistance to a range of abiotic and biotic stresses, consumer-oriented quality traits.
3. Distribution of international nurseries particularly targeted to a number of main agroecosystems, via national wheat programs across the world.
4. Analysis of international yield tests and free access to all amassed data.
5. Global disease and pest monitoring to ensure significance of current local, global and regional breeding activities.
6. Capacity building as well as training of research partners.
7. Regular contact among research partners through workshops, consultation, etc., to help classify the trending technology needs.

Through the IWIN, wheat germplasm has spread persistently since the Green Revolution and is now not only tremendously well represented in farmers' fields of the developing world, but is also frequently realized in the pedigrees of wheat lines in developed countries (Braun et al., 2010).

The continued efficiency of these wheat breeding strategies have been confirmed right up the present with yield data from hundreds of testing sites across the world showing average genetic yield gains of 1% every year in germplasm targeted to water restricted environments (Manes et al., 2012), and 0.6% per year in fully irrigated environments (Sharma et al., 2012) based on lines distributed since 1995. Germplasm distributed by IWIN is sustained resistance to the full range of diseases that normally affect wheat (Braun et al., 2010).

As an outcome, resource-poor farmers in LDCs are defended financially from having to apply pricey fungicide, while the environment is safeguarded from the additional agrochemical burden. A significant spin-off of the IWIN are huge phenotypic datasets that have permitted breeders to recognize germplasm with either explicit adaptation to local diseases and challenges, or broad (spatial and/or temporal) acclimatization to many locations as well as cropping systems (e.g., Gourdji et al., 2012).

In spite of this achievement, genetic gains still fall short of accommodating the projected demands by the year 2050 (Rosegrant and Agcaoili 2010), a mismatch that signifies a serious challenge for future food security, particularly in the light of the challenges related with climate change.

One of the most effective ways to tackle this challenge from the perspective of genetic improvement is through a more systematic use of genetic resources in breeding. To accomplish this necessitates a series of steps that are sketched in the next few sections of this chapter.

7.3.1. Identify Crop Characteristics Conferring Stress Adaptation

Many crop types have been stated as conferring enhanced stress adaptation (e.g., Rebetzke et al., 2009; Cossani and Reynolds 2012) and well-ordered phenotypic studies can suggests whether these traits may be balancing in distinctive combinations (Reynolds et al., 2007a).

Although, the only way to validate their value finally is to find sources with decent expression of the traits and into elite backgrounds. The tactic has been quite successful within the IWIN for creating a new generation of drought improved lines, as will be discussed in the last section.

Some of the traits that have showed to be most beneficial for both drought and heat adaption comprises (*see,* Richards 2006; Rebetzke et al., 2009; Reynolds et al., 2010):

1. Cooler canopy temperature that suggests entree to subsoil water under drought and an acceptable vascular system – comprising roots – to balance evaporative demand at high vapor pressure shortfall archetypal of heat stress in dry environments.
2. Rapid early ground cover to evade extravagant evaporation of water at the soil surface back to the atmosphere.
3. Intrinsic transpiration efficacy that makes sure traditional use of water when it is comparatively ample as a water planning strategy.
4. Epicuticular leaf wax that imitates superfluous radiation and lowers evaporation from the leaf surface, thus decreasing the risk of dehydration and photo-inhibition, correspondingly.

5. Remobilization and accumulation of water-soluble carbohydrates in storage organs such as stems allowing grain-filling to endure even when postanthesis stress is too stern to allow satisfactory carbon acclimatization.
6. Membrane thermo-stability that can be assessed using electrolyte leakage, and chlorophyll fluorescence that, though it has not been used methodically in choosing parents, has become a fresh attention for screening.

7.3.2. Exploration of Genetic Resources for Adaptive Traits: Landraces

CIMMYT and other wheat breeding programs have been making use of landraces to widen the wheat gene pool for several decades (Smale et al., 2002), mainly for disease resistance traits as these are comparatively cool to identify in un-adapted material and choose for in separating progeny (Reynolds and Borlaug 2006).

More recently Mexican landraces were collected -after roughly 500 years of essentially natural selection in some of Mexico's harshest rain-fed environments.

These were inspected for drought and heat adaptive traits, and a few of the lines show excellent characteristics such as deep roots and potential to store water-soluble carbohydrates in the stem (Reynolds et al., 2007b). Recently, over 70,000 accessions of the World Wheat Collection have been inspected for adaptation to drought and heat stress -typically for the first time-, and diverse panels have been gathered for genetic analysis.

These comprises of bread wheat and elite durum lines, landraces from dry and hot regions across the world, and lines derived from interspecific hybridization with wheat's lineages, embracing 're-synthesized' wheat.

Other more targeted methods comprise characterization of panels recognized using agro-ecological data, also known as the Focused Identification of Germplasm Strategy or FIGS.

The opinion is that landraces from drought and heat-stressed regions are more probable to comprise stress adaptive traits. In the case of drought, we attained 204 landraces deriving from drought-affected areas.

These were grown in Sonora, Cd Obregon, under drought conditions (only one irrigation at sowing offering roughly 180 mm of available water to 1.2 m depth, and no noteworthy rainfall during the growing season).

CIMMYT varieties were comprised as checks. Total above-ground biomass was projected and grain yield after separating.

7.3.3. Inter-Specific Hybridization to Broaden the Crop Gene Pool

Although, it is important to note that the gene pools used in traditional breeding are comparatively limited (Hajjar and Hodgkin 2007), a vast pool of genetic resources is obtainable for breeding. Different gene pools have been described contingent on the effort of employing them.

One of the best ways to use are those from the primary gene pool signified by germplasm that distribute a common genome but which have become isolated from mainstream gene pools such as landraces.

The secondary gene pool is signified by closely related genomes that can be employed through interspecific hybridization, and would comprise the growth of so-called "re-synthesized" or "synthetic" wheat, where tetraploid durum wheat has been hybridized with *Aegilops tauschii*, the ancestral donor of the D-genome, to reconstruct hexaploid bread wheat (Mujeeb-Kazi et al., 1996).

This tactic has been quite successful in presenting disease resistance and drought adaptive traits (Reynolds et al., 2007b; Trethowan and Mujeeb-Kazi 2008).

Several thousands of accessions of both ancestral genomes (AB and D) persist as candidates for interspecific hybridization. Physiological classification is being used to assists in choosing the most promising accessions, in addition to selection among the initial products (primary 'synthetics'), for genetic and prebreeding studies.

In parallel to this, genotyping of prevailing potential progenitors and primary synthetics (AB tetraploid, and *Ae. tauschii*) is being carried out to target under-used genetic diversity.

Engaging in interpretation of impacts from using synthetic wheat in breeding have shown substantial contributions to drought adaptation (Lopes and Reynolds 2011), and it seems that the presentation of alleles from the wild D genome have amplified the potential of roots to partition resources to lesser soil profiles when suffering water stress (Reynolds et al., 2007b).

A current interpretation looked at the similar lines under heat stress in the hot desert of NW Mexico. To create modest and more extreme heat stress in the field, fully irrigated crops were planted later than normal in the months of February and March, correspondingly (majority of the time sowing happens in the months December).

Results showed highly imperative impacts of synthetic derived (SYN-DER) lines in hot, irrigated environments on both biomass and yield.

7.3.4. Transgenic Approaches

It is worth noticing that transgenic approaches eliminate all taxonomic limits to plant improvement. The control of genes through biotechnology has offered the prospect of genetic engineering plant responses to abiotic stresses such as drought, heat, and salinity (Qin et al., 2011).

The regulated expression of stress-persuaded transcription factors, for instance, has been identified as an appealing tool for enhancing stress tolerance, since transcription factors can govern the expression of a large number of pertinent downstream genes correlated to abiotic stress responses in genetically adapted plants (Nakashima et al., 2009).

Transcription factors such as DREB2, DREB1/CBF, AREB/ABF, and NAC, are used to enhance stress tolerance to abiotic stresses in several grasses comprising rice and wheat.

Recent efforts to test the functionality of several transcription factors comprised in multifaceted physiological responses and to assess the effect of these under either inducible or constitutive promoters have been made in wheat.

Transgenic wheat lines were advanced at CIMMYT using new concepts generated at the Rikagaku Kenkyūjo (RIKEN) and Japan International Research Center for Agricultural Sciences (JIRCAS). The resulting lines were then portrayed under glasshouse conditions and open field trials directed at CIMMYT's subtropical experimental station in Tlaltizapán, Morelos, Mexico (18° 41′ N, 99° 10′ W, 940 m asl).

In these open-field trials, grain yield performance and physiological traits of 116 homozygous low copy number GM wheat lines (14 gene-promoter combinations) were assessed under distinctive water regimes. The lines assessed in the field showed no pleiotropic effect, nor erratic unwanted effects in comparison to control lines (Saint Pierre et al., 2012).

On the basis of 3-year analysis, the most favorable lines exhibited an average increase of yield under drought of 10% when rivaled with nontransformed line (var. Fielder).

Even though, there may be witness of promising results, several challenges have yet to be address before foreseeing a high impact of genetically adapted lines on wheat production.

There is a need for greater effort to tackle the identification of adequate gene and gene-promoter combinations, field selection strategies, and the introduction of the suitable transgene in adequate backgrounds.

Regulation processes, marketing, and commercialization of genetically modified products are peer challenges to face. From the perspective of the technical side, an imperative step in the case of genetically modified wheat is that more effective alteration protocols as well as exact expression systems would require to be stated.

The wheat variety Fielder is normally used in alteration because of its noble regeneration efficiency, and embryogenesis capacity. Though, it has lesser yield potential in comparison to modern elite wheat lines.

Then, after incorporation of the transgenes in wheat backgrounds (Fielder), it is essential to transfer them to modern elite lines, which is presently attained by backcrossing. Predictions on the basis of extrapolating yield enhances from the transmuted line Fielder to new high performing germplasm (elite lines) should be made with attention.

The challenge is, thus, to halt the yield barriers in contemporary elite wheat lines to ultimately attain a noteworthy improvement in wheat grain yield under stress. Cooperative and collective interventions from molecular biologists, breeders and physiologists, are needed.

7.3.5. High Throughput and Precision Phenotyping

The effective examination of genetic resources -for crop improvement and to identify mechanistic and genetic bases- necessitates a combination of accuracy and high throughput phenotyping tactics. Non-intrusive high spatial resolution spectral imagery can be employed to the examining of physiological characteristics such as hydration status, canopy temperature, and pigment composition, as well as letting estimates of agronomic traits such as yield and biomass. Spaceborne remote sensing platforms have proven effective at computing some of these features at a field scale, though their spatial resolution proves too low for precise data recovery at plot level in a plant breeding context. While ground-based remote sensing is used for projecting agronomic and physiological traits at a plot scale, temporal deviations of environmental variables, like air temperature, can initiate confounding factors, particularly when applied to large trials.

Low-level airborne remote sensing platforms address these limitations, allowing for rapid, nondestructive screening of plant physiological properties over sizeable areas, with adequate resolution to attain information at plot level while being able to assess thousands of genetic resources in the field with one take.

Rapid enhancements in remote sensing technologies, instruments availability and data processing has made it simpler to apply remote sensing techniques to research several plant properties (e.g., Leinonen and Jones 2004; Jones et al., 2007; Berni et al., 2009).

The surge in demand for large-scale vegetation checking means that there is a move to such remote sensing applications in which instantaneous measurements of huger target areas can be made readily. In a recent study, Tattaris et al. (2013) applied a low-level airborne remote sensing platform to originate indices linking to plant properties such as water status, canopy temperature, and pigment composition.

Sampling was executed with a thermal camera and multispectral camera mounted on helium-filled tethered blimp and an eight-rotor unmanned aerial vehicle (UAV). Airborne indices were authorized by similar indices collected at ground level.

These ground-based measurements have already been showed to be associated with biomass and yield (e.g., Reynolds et al., 1994; Aparicio et al., 2000). This strong agreement between approaches acts to authenticate the use of the airborne indices.

Moreover, noteworthy genetic correlations were found between the airborne indices derived using yield/biomass and imagery, larger than the corresponding correlations between yield/biomass and the ground-based measurements.

This is possibly because airborne tactics lessen error in two ways. Firstly, instantaneous measurement allowable through airborne platforms evade the perplexing effects concerned with environmental drift when computing plots one at a time with ground-based instruments.

Secondly, aerial imaging techniques open the pathways for data smoothing through removal of outlier pixels. These factors give added confidence to make use of these methodologies in screening genetic resources and breeding progeny on a large scale where there is involvement of thousands of plots.

7.3.6. Strategic Crossing to Achieve Cumulative Gene Action

Trait selection has made persistent development in wheat breeding through integrating agronomic traits such as flowering and height time, quality parameters derived by end-use, defiance to a spectrum of prevailing diseases, and yield on the basis of multilocation trials (Braun et al., 2010).

Though, current genetic gains are not enough to accommodate future demand while climate change portends to erode these further. Breeding more precisely for drought and heat adaption can assists in increasing genetic gains (Reynolds et al., 2009). The primary objective of strategic trait-based crossing is to accrue traits that will be balancing for a given target environment (target-traits). Under water-limited situations, traits that advance water use efficiency, water uptake, and partitioning to yield, correspondingly, are likely to work synergistically to surge overall productivity (Richards 2006). Attaining this in a practical breeding context comprises of the following interventions, expressed chronologically:

1. Target traits are recognized using conceptual models as labeled above.
2. Genetic resources are separated to offer potential sources of target traits.
3. Candidate parents are illustrated for traits that may assists in genetic gains in target environments.

7.4. WHEAT GENETIC RESOURCES EVOLUTION AND THEIR ROLE IN BREEDING

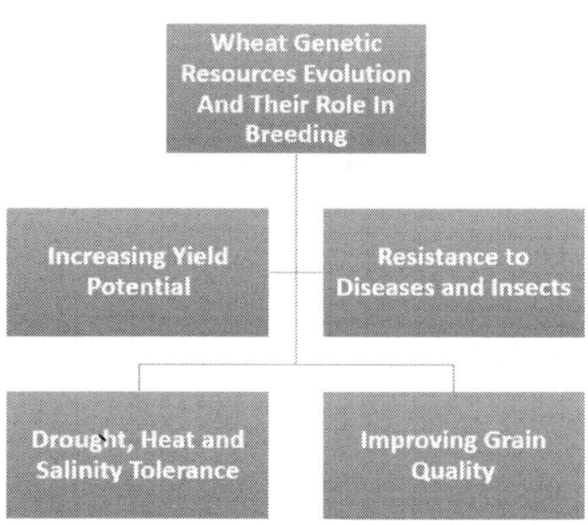

Figure 7.3. Wheat genetic resources evolution and their role in breeding.

Based on genomic constitution, evolutionary distance from each other and intercross ability, genetic resources have been grouped into primary, secondary as well as tertiary gene pools. The primary gene pool is composed of early domesticates, wheat landraces and wild species. In addition, they are directly crossable to bread as well as durum wheat.

The secondary gene pool comprises of polyploidy Triticum in addition to Aegilops species that share not less than solitary homologous genome with the cultivated types. This gene pool consists of, for example, the diploid S-genome (similar to the B genome) and T. timopheevii (AAGG) species from Aegilops.

Gene introgression of this group into cultivated wheat is challenging in addition, embryo rescue is often essential to attain F_1 hybrids. The tertiary gene pool mostly involves polyploids as well as perennial diploids.

It has no mutual genome constitutions regarding the cultivated species, comprising species belonging to other genera, for instance, Secale, Amblyopyrum, Dasypyrum, Agropyron, Thinopyrum, Elymus, Eremopyrum, Hordeu, Elytrigia as well as Leymus. Gene transfer is extremely complex in nature, and in order to enhance translocations, it requires special techniques such as irradiation or gametocidal chromosomes and embryo rescue.

Germplasm collection has been carried out across different countries for breeding and breeding purposes with the chief objective of recognizing noble traits for resistance to biotic as well as abiotic stresses. Nevertheless, of all the germplasm collections deposited in gene banks, breeders have used only 10% of the materials. There are a number of reasons that are connected with this low-level consumption of gene bank resources.

The utmost essential reasons are: materials in the gene bank are obsolete and clumsy, difficult to cross, poorly defined as well as characterized, and time consuming to transfer genes of interest.

In addition, breeders argue that they possess a variety of traits in the elite lines of their programs related to breeding. Nevertheless, the worth of genetic resources should be evaluated not only by the authentic rate of usage but for their ability to deliver sources of novel diversity continuously for most of breeders sought traits, and these are projected to be even more significant for refining complex traits together with climate change adaptive traits. The prevailing literature displays that wheat was among the few crops for which genes are transmitted proficiently from their wild relatives. The role of conserved wild relatives of landraces as well as wheat is projected to rise with new expansion of tools letting efficient pyramiding and introgression of genes and efficient selection as well as follow up of the same targeted by the crosses comprising usage of genotyping together with sequencing information.

Better targeting the traits in the existing collections through applying best subsetting approaches alternative to random sampling or core collections can further enhance the use of wheat genetic resources. In order to select best bet subsets for specific traits sought by breeding programs, the Focused Identification Germplasm Strategy is developed in this regard. FIGS approach stimulates natural selection and uses filtering of algorithms to define the relations regarding conditions of environment and the sought trait.

The importance of this approach has enabled the identification of novel diversity for disease resistance as well as insect resistance in wheat. However, genetic resources have played a major role in the following key areas of wheat breeding.

7.4.1. Increasing Yield Potential

In the nineteenth century, the first scientific breeders started selecting as well as crossing landraces and cultivars from genetic pools and diverse origins for obtaining the combinations that led to superior varieties. In the first decades of the twentieth century, Nazareno Strampelli, who was the Italian breeder used the Japanese variety Akakomugi in his crossing block and made crosses with Italian landraces and breeding lines.

The most successful varieties were the results of such crosses, for instance, Mentana and Ardito possessing *Ppd-D1* alleles for photoperiod insensitivity as well as *Rht8c* for short straw. Later on, these varieties became the backbone of most of the new varieties that were developed in the Mediterranean countries, China, USSR in addition to some South American countries.

In the 1950 s, the Nobel laureate Norman E. Borlaug has made the Norin-10/Brevor cross that presented the *Rht-D1* as well as *Rht-B1* dwarfing alleles that resulted in the development and release of input responsive semidwarf varieties of the "Green Revolution." These dwarfing genes related to Norin-10 were originally obtained from the Japanese landrace "*Shiro Daruma*."

In spite of the fact that these dwarfing genes have contributed significantly to nutrient use efficiency, input responsiveness and increase yield, it is important to note that their introgression was very difficult and time consuming as well. At ICARDA and CIMMYT, elite lines × synthetic wheat and spring × winter wheat crosses are intensively used for developing high yielding wheat genotypes with resistance and broad adaptation to the key biotic as well as abiotic stresses in the face of climate change.

7.4.2. Drought, Heat, and Salinity Tolerance

Wheat landraces have been exploited for tolerance to several abiotic and resistance to main biotic stress. In addition, they are the sources of numerous significant genes comprising those connected with drought tolerance.

It has been observed that recent wheat lines with synthetic derived wheat (SDW) or wild relatives in their background have comparatively better performance across different environments for diverse stresses such as drought, salinity, and heat. Under heat and drought conditions, synthetic hexaploid wheats (SHW), and synthetic derived wheat (SDW) of CIMMYT have exposed superior performance for grain yield. In the same way, better water use efficiency, high root biomass and drought tolerance was reported from translocation lines carrying7A/7DL from *Agropyron elongatum*.

Varieties of wheat-derived from synthetic wheat have revealed 11%–30% higher grain yield under rain-fed conditions in southern as well as northern Australia, representing the significance of the synthetic wheat for drought tolerance. Lage and Trethowan conducted multilocation data analysis of the semi-Arid wheat yield trials of CIMMYT that indicated the synthetic derived wheat genotype "Vorobey" was the top-yielding across most of the testing locations worldwide.

Synthetic derived wheat has been also reported to have tolerance or resistance to salinity in addition to boron toxicity that indicates the significance of wild relatives in refining common wheat. Wheat wild relatives are expected to hold more novel genes for adaptation to cold, drought, salinity, heat, frost, acidity stresses, most of which will be improved with the change in climate.

7.4.3. Resistance to Diseases and Insects

Genetic resources have played an important role in the identification of essential genes and thereby assisted in reducing losses in quality as well as yield because of insects besides major diseases. Resistance sources are often obtained from the centers of origin either from the wild relatives or from landraces due to the coevolution of the host as well as the pathogen.

The *Sr2* gene for stem rust resistance is one of the good examples of the contribution of landraces to disease resistance. It was originated from the emmer wheat cultivar "Yaroslav" and in the year 1923, Dr. McFadden was able to fruitfully unify it into bread wheat by conventional hybridization. The *Sr2* gene deployed in combination with many other major and minor stem rust genes for almost more than 100 years. It significantly contributed in controlling stem rust of wheat across several nations.

However, its effectiveness was overcome with the identification of the new races since 1999 including UG99. Sr24 from *Thinopyrum ponticum* (1RS.1AL), Sr36 from *Triticum timopheevi* and Sr38 from *Triticum ventricosa* are other successful resistance genes for stem rust. UG99 and its derivatives broke most of the stem rust resistance genes. Sr22, Sr25, Sr26, Sr39, Sr42 and Sr45 are some of the genes that are still effective.

In the same way, the synthetic derived wheat has contributed resistance sources to many insects (Russian wheat aphid, Hessian fly and Sun pest) as well as several other diseases (leaf rust, septoria, yellow rust, tan spot, and nematodes).

7.4.4. Improving Grain Quality

The increased number of countries that apply price premiums for high quality wheat is changing the trend while quality has not been a breeding driving force in many countries in the world. Actually, farmers can get up to 10% more revenues on their wheat by just increasing content of protein by 1.5% in countries such as the United States.

These premiums increase in unfavorable years, like in year 2011, when wheat quality premiums rose up to 25% when farmers of Canada increased the content of protein by 2%. Genetic resources like wild relatives, landraces in addition to synthetic wheat have been reported to contribute to improving the quality of wheat grain. Landraces are considered the desired source of alleles for wheat breeding besides modern improved wheat lines.

This is because of their genetic diversity as well as their proximity to modern cultivated wheat in comparison to several other sources like synthetics. Large diversity of landrace makes them source of bread making related alleles such as bread making related alleles such as alternative prolamine alleles. Moreover, as proven by the high levels of micronutrients that are found in a collection of Sicilian landraces, they can also be sources of biofortification for wheat.

Nevertheless, hexaploid wheat is an outcome of a very restricted number of unique hybridization events among the wheat genome contributors. As a result, the original genetic diversity of early hexaploidy wheat and so of modern wheat is restricted. The reformation of the unique hybridization events with new contributors of the A, B, and D genomes to increase wheat diversity as well as quality is the main solution for this genetic bottleneck. Synthetic wheat that is developed by crossing of emmer wheat with *Aegilops tauschii* followed by chromosome doubling have exhibited semihard to hard grain texture as well as incredible variations for SDS-S values, protein content, kernel weight together with plumpness. Lage et al., stated that synthetic hexaploids have an average protein value of 15.5% as compared to the common bread wheat cultivars like "Seri M82" that have on average 13.1% protein. In the same way, the synthetic wheats have higher SDS values as well as longer grains. Protein content among synthetics exhibited no correlation with SDS values but significantly negative correlations with grain weight and plumpness. Without being affected by these negative correlations, Lage et al., have recognized 10 superiors synthetic hexaploid wheat created from 9 diverse emmer parents with high values of SDS, plumpness, protein content and kernel weight that indicate genetic variation for quality traits from tetraploid emmer wheat can be easily transmitted into hexaploid common wheat. Genotypes of spring bread wheat together with synthetic wheat in their background have been reported with high content levels of Fe and Zn. In most of the developing countries, speedy arrangement of such varieties of wheat would help to develop nutritional security.

7.5. BREEDING METHODS AND APPROACHES TO INCREASE GENETIC GAINS

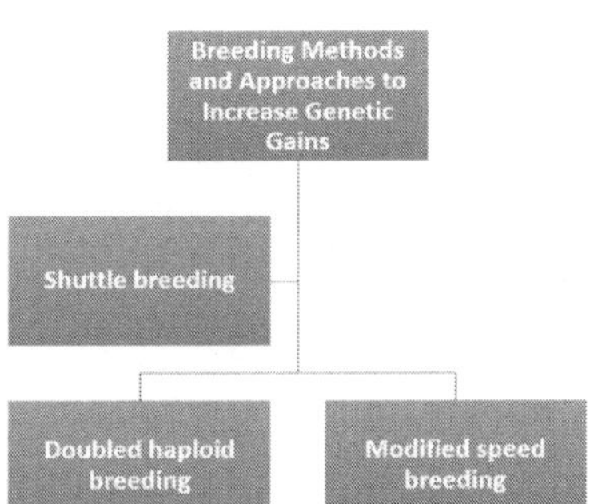

Figure 7.4. Breeding Methods and Approaches to Increase Genetic Gains.

In the face of climate change with continued water shortage both in irrigated and rained environments, it is important to develop high yielding wheat genotypes with high water use efficiency, which combines both high yield potential and resistance to drought and heat stresses along with resistance to the major biotic stresses.

To this end, it is important to formulate and deploy efficient breeding methods and strategies that enable to increase genetic gain while using the available resources efficiently. The first and most important thing is to define the target regions (mega-environments) and identify the corresponding production constraints.

This is followed by assembling crossing blocks targeting the major mega environments. At ICARDA, we use shuttle breeding, doubled haploids, marker-assisted selection, key location phenotyping, hot spot screening, international nurseries distribution and partnership with the NARS.

It is essential to develop high yielding wheat genotypes with high water use efficiency that combines both high yield potential and resistance to drought and heat stresses along with resistance to the major biotic stresses in the face of climate change with continued water shortage both in rained as well as irrigated environments.

Under the circumstances, it is essential to formulate as well as deploy efficient breeding strategies along with methods that enable to increase genetic gain while using the available resources efficiently. Defining the target regions (mega-environments) and identifying the corresponding production constraints is the first and most important thing.

This is followed by assembling crossing blocks that target the major regions or mega environments. We use doubled haploids, shuttle breeding, marker-assisted selection, hot spot screening, key location phenotyping, international nurseries distribution and partnership with the NARS at ICARDA.

7.5.1. Shuttle Breeding

Shuttle breeding is a result-oriented crop improvement program in which the same species of crop is grown in different contrast conditions, i.e., two or more diverse ecological conditions to advance the generations and shorten the breeding cycle.

It was originally used at the International Maize and Wheat Improvement Center (CIMMYT) by Norman E. Borlaug using the Obregon station a winter season irrigated dry location at 27.5° N, 40 m.a.s.l. in the state of Sonora, in Mexico and Toluca station (a summer season location located at 27.5° N, 40 m.a.s.l. with 800–1000 mm average rainfall) in the state of Mexico. These two stations are very different in respect of altitude, latitude and rainfall.

These stations have proven to be effective for introducing photoperiod insensitive genes in spring bread wheat. The photoperiod insensitive genes (*Ppd1* as well as *Ppd2*) together with the dwarfing genes (*Rht1* and *Rht2*) have facilitated the development of widely adapted, lodging resistant, high yielding and input responsive varieties of spring bread wheat. Since 2008, the CIMMYT wheat-breeding program has established the Njoro station in Kenya as an additional shuttle breeding location to screen germplasm for stem rust resistance (Ug99 together with its variants).

In the same way, at ICARDA, the spring bread wheat breeding program uses a shuttle breeding program for the winter × spring crosses including Terbol station (34° N; 36° E, 900 m.a.s.l) in Lebanon (winter × spring crosses at the time of winter season, F_1 during the summer season), evaluation of F_2 populations during the winter season at Sids station (29° N; 31° E, 32.2 m.a.s.l.)

In Egypt, evaluation of F_3 at the time of summer season at Kulumsa (08° N; 39° E, 2220 m.a.s.l) in Ethiopia and evaluation of F_4 populations during the winter cycle at Merchouch station (33.6° N; 6.7° W, 430 m.a.s.l) in Morocco.

Shuttle breeding facilitates additional recombination events, selection of transgressive segregants, and selection of resistance against biotic as well as abiotic stresses with broad adaptation in addition to the cycle shortening. Such shuttle breeding in winter as well as summer season is possible in case of spring bread wheat; however there were determinations to carry out shuttle breeding for winter wheat between the summer station of CIMMYT at Toluca, Mexico and Oregon state university, USA. Winter wheat is grown during the winter cycle because it normally requires vernalization, and therefore it is possible dominantly one generation per year advancement under condition of field.

7.5.2. Doubled Haploid Breeding

A doubled haploid is a genotype formed when cells of haploid (n) undergo chromosome doubling. Artificial production of doubled haploids is important in plant breeding. *In vitro* haploid production using microspore culture, embryo culture or anther culture using wheat × maize crosses followed by chromosome doubling prominently increases the precision and efficiency of the selection process in wheat breeding and develops the production of homozygous wheat lines in a single generation.

It also facilitates estimation of genetic variance and the number of genes for quantitative traits, detection of linkage and gene interactions. Doubled haploids facilitate genetic transformation and mutation studies. It also enables detection as well as identification of substitutions, translocations and chromosome addition lines. Numerous varieties issued from DH breeding strategies have been released in the last few decades in many countries such as Hungary, China, Romania, France, and Canada.

Above and beyond variety development, wheat DH populations have also been used in identifying quantitative trait loci and to unveil their mechanisms of action and creation of molecular marker maps. New wheat QTL for agronomic performance, end-use quality-related traits, and disease resistance has been recognized with the usage of DH populations.

In studies relating genotypes allotted from SSD, DH and pedigree methods, no substantial variances in grain yield were found among several populations while the parental varieties were closely associated in their pedigrees.

It is worth noticing that the use of doubled haploids in breeding has two crucial benefits. On the one hand, DH assists in shortening the cycle in wheat by 16 to 50% depending on the growth habit (winter or spring) and the production from which the DH are produced (F_1, F_2, BC_1F_1, etc.).

In a usual breeding strategy, only after at least 6 generations, the level of homozygosity is large enough to commence genotype screening and initial yield trials.

In winter wheat, it is usually seen that only one generation per year can be attained under field conditions, thus the use of doubled haploids can curtail from 8 to 4 years the time needed to achieve preliminary yield trial when the expanded haploids are formed from F_1 plants.

In spring wheat, the benefit of DH in comparison to normal breeding is residual because of the potential of growing two generations per year. On the one hand, DH plants are fully homozygous that enhances the efficacy of selection for both quantitative and qualitative traits since recessive alleles are fixed in one generation and unswervingly expressed. In some cases, such as mutation breeding, the benefit is even larger, particularly for quantitative traits. Strategies with the objective to selection accurateness in DH breeding have been efficaciously used across the world. Mainly, the surge in correctness can be attained at the cost of the cycle length by deferring the generation from which the DH are formed. Backcross-DH strategy has been one of the pertinent examples.

Thus, in Iran, yellow rust resistance was successfully and competently transferred to the adapted receptor cultivar "Ghods" by forming DH from tested BC_2F_2 plants. The DH plants formed exhibits similar yield and phenology as the adapted parent but were resistant to two isolates of yellow rust and were produced in a briefer period of time.

Although, if the number of traits to introgress, their amount or heritability of sample needed to assess them (i.e., quality tests) can sternly lessen the efficacy of this method.

7.5.3. Modified Speed Breeding

Speed breeding' technology can be defined as the technology for production of up to six generations of wheat per year comprising extended duration of light and early harvesting of wheat under glasshouse conditions.

This effectively helps in shortening the breeding cycle through acceleration of generation advancement. Though, it might be confined in its scope to screen the population for distinctive abiotic and biotic stresses along with the challenge of handling of large number of populations needed at the F_2 stage.

For example, ICARDA, developed a modified speed breeding for elite × elite crosses where by crosses are ready in the plastic houses during winter cycle accompanied by production of F_1 in the glasshouse using speed breeding in spring followed by production of F_2 in summer under field condition by making use of irrigation at Merchouch station where population is being selected on the basis of selected bulk scheme for biotic and abiotic stresses.

The F_3 population is developed during the winter cycle in the field at Merchouch (December–June) where individual plants/ spikes are selected on the basis of pedigree selection scheme and fostered to head-rows during the summer cycle. Initial yield trials are cultivated during the winter season across central locations using the selected head-rows.

Selected lines are updated into first year and second-year Advanced Yield Trials (AYTs). The second-year AYTs are also increased at a similar time in large plots of six rows with 0.2 m spacing between them and 20 m length. Selected genotypes on the basis of grain quality and agronomic data interpretation are composed into the distinctive international nurseries and yield trials and dispersed globally on the basis of request by the national programs.

The top-performing genotypes are also comprised in the crossing block, making the entire breeding cycle finished in 4 years. Such a method is greatly efficient to enhance genetic gain through abridging the breeding cycle and permitting selection and generation advancement under abiotic and biotic stresses dominant during winter and summer seasons.

7.6. APPROACHES TO COMBAT CLIMATE CHANGES

Figure 7.5. Extreme climatic conditions.

Source: Image by Eos.org.

It is generally seen that variation in the environment has a long-lasting impact on agriculture and food security across the world. Food security and safety are susceptible by the extreme weather conditions, and it is not a contemporary problem.

But formerly, no effort was undertaken to address this problem. Thus, to cope with such weather deviations is one of the most critical problem that need to be addressed globally. In order to make sure that crops be adapt to changing environmental stresses, some of the approaches that need to be taken into consideration are discussed as follows.

7.6.1. Cultural Methodologies

Recently, some experiments stated investigations of the strategies trialed by farmers to address the climatic variation for plant adaptation. There are several optimal approaches that can be adopted by farmers, including abiotic factors such as harvesting time and altering planting, a collection of crops with short life cycles, irrigation techniques, crop rotation, and variation in cropping schemes.

Under climatic stress conditions, all of these tactics are extremely advantageous for crop adaptability. Modification in sowing time, the cultivation of novel crops and application of drought-resistant cultivars are of the imperative strategies to curtail the climatic variability danger and offer better adaptability to crop plants for assuring food security and safety. Another plant adaptability approach is by means of crop-management techniques that have the potential to augment crop development under several environmental stresses. The choice of sowing time, optimum irrigation practices, and planting density are critical techniques to address weather stresses.

Fertilizers are also extremely important to lessen global warming effect and support the plant for better flexibility. It offers ample energy to plants and is favorable to maintain the soil fertility and increase productivity. Thus, the significance of fertilizer in nurturing the world is irrefutable.

7.6.2. Conventional Techniques

Under several environmental stresses, plant breeding exhibits dynamic techniques in crop development and betterment. It gives a way to potentially assure food safety and security under extreme weather variations and help plants drip from several stresses through a critical phase of plant growth by forming stress-resistant cultivars.

Genetic divergence interpretation is basically used for inbreeding, polymorphism, assessment, recombination and assortment to achieve plant perfection, and is among the chief aspects for stating accomplished inbreeding. Genetic divergence analysis is believed to be an imperative method for the development of new cultivars on the basis of genetic distance and similarities. For genetic studies, landraces are an imperative source, for instance, a wheat landrace kept in data bank includes bigger genetic variance and is a respected basis for stress resistance as it comprises cultivars adaptable to diverse environmental stress.

7.7. CONCLUSION

At the end, it is concluded that there are several breeding strategies available and the selection of best strategies that can adapt to the local environment depends on a number of factors. Breeding strategies are responsible for improving the overall productivity by tackling the various stressors. Local environment plays a very important role and the overall efficiency depends upon the strategy that is choosing by taking into consideration the prevailing environment and other factors.

Climate change is another factor that should be in mind of the farmers while framing breeding strategies, as following an identical strategy for over a period of time may loose its efficiency because of change in climate.

With the increase in population, the demand of wheat will be continually increase, therefore it is highly important to not compromise with the quality of wheat along with its productivity.

REFERENCES

1. Ahmed, I., Ullah, A., Ur Rahman, M., Ahmad, B., Wajid, S., Ahmad, A., & Ahmed, S., (2019). Climate change impacts and adaptation strategies for agronomic crops. *Climate Change and Agriculture.*
2. Anwar, M., Liu, D., Macadam, I., & Kelly, G., (2012). Adapting agriculture to climate change: A review. *Theoretical and Applied Climatology, 113*(1, 2), 225–245.
3. Cereal Foods World, (2020). *Breeding Approaches for Nutritional Quality Traits in Wheat, 65*(3).
4. Chakraborty, S., & Newton, A., (2011). Climate change, plant diseases and food security: An overview. *Plant Pathology, 60*(1), 2–14.
5. *Crop Breeding, Genetics and Genomics*, (2019). *Genetic Gains in Wheat Breeding and Its Role in Feeding the World.* [online] Available at: https://cbgg.hapres.com/htmls/CBGG_1052_Detail.html (accessed on 7 July 2021).
6. Daloz, A., Rydsaa, J., Hodnebrog, Ø., Sillmann, J., Van, O. B., Mohr, C., Agrawal, M., et al. (2021). Direct and indirect impacts of climate change on wheat yield in the Indo-Gangetic plain in India. *Journal of Agriculture and Food Research, 4*, 100132.
7. Dreccer, M., Bonnett, D., & Lafarge, T., (2013). Plant breeding plant breeding under a changing climate plant breeding under a changing climate. *Sustainable Food Production*, 1296–1307.
8. Gazal, A., (2021). *Breeding Climate Change Resilient Maize and Wheat for Food Security*. [online] www.researchgate.net/. Available at: https://www.researchgate.net/publication/339974623_Breeding_climate_change_resilient_maize_and_wheat_for_food_security (accessed on 7 July 2021).
9. Geng, X., Wang, F., Ren, W., & Hao, Z., (2019). Climate change impacts on winter wheat yield in Northern China. *Advances in Meteorology, 2019*, 1–12.
10. Ibrahim, M., Ouda, S., Taha, A., El Afandi, G., & Eid, S., (2012). *Water Management for Wheat Grown in Sandy Soil Under Climate Change Conditions.* [online] scielo.conicyt.c. Available at: https://scielo.conicyt.cl/scielo.php?script=sci_arttext&pid=S0718-95162012000200001 (accessed on 7 July 2021).
11. Kajla, M., Yadav, V., Khokhar, J., Singh, S., Chhokar, R., Meena, R., & Sharma, R., (2015). Increase in wheat production through management of abiotic stresses: A review. *Journal of Applied and Natural Science, 7*(2), 1070–1080.

12. Raza, A., Razzaq, A., Mehmood, S., Zou, X., Zhang, X., Lv, Y., & Xu, J., (2019). Impact of climate change on crops adaptation and strategies to tackle its outcome: A review. *Plants, 8*(2), 34.
13. Reynolds, M., Ortiz-Monasterio, J., & McNab, A., (2021). *Application of Physiology in Wheat Breeding*. [e-Book] CIMMYT. Available at: https://repository.cimmyt.org/xmlui/bitstream/handle/10883/1248/74619.pdf (accessed on 7 July 2021).
14. Reynolds, M., Tattaris, M., Cossani, C., Ellis, M., Yamaguchi-Shinozaki, K., & Pierre, C., (2015). Exploring genetic resources to increase adaptation of wheat to climate change. *Advances in Wheat Genetics: From Genome to Field*, 355–368.
15. Shew, A., Tack, J., Nalley, L., & Chaminuka, P., (2021). *Yield Reduction Under Climate Warming Varies Among Wheat Cultivars in South Africa*. [online] www.nature.com. Available at: https://www.nature.com/articles/s41467-020-18317-8 (accessed on 7 July 2021).
16. Snowdon, R., Wittkop, B., Chen, T., & Stahl, A., (2020). Crop adaptation to climate change as a consequence of long-term breeding. *Theoretical and Applied Genetics*.
17. World-agriculture.net. (2021). *Using Climate Information to Support Crop Breeding Decisions and Adaptation in Agriculture* | World Agriculture. [online] Available at: http://www.world-agriculture.net/article/using-climate-information-to-support-crop-breeding-decisions-and-adaptation-in-agriculture (accessed on 7 July 2021).

Chapter 8

Crop Adaptation to Climate Change

CONTENTS

The chapter on crop adaptation to climate change explains the various aspects of climate change that influences the crop as well as how crop adaptation comes in the play in order to address the climate change. This chapter also sheds light on the habituating agriculture to the sudden changes in climate.

This chapter also explains the various aspects of sudden changes in the climate, their impacts along with the various adaptation strategies for agronomic crops. This chapter also provides highlights on how will climate change influences the cropping system. This chapter also mentions how people can acclimate crops and cropping systems to climate change.

8.1. INTRODUCTION

Climate change is a solemn threat to agriculture and food security. Extreme weather conditions and transmuting patterns of precipitation lead to a decrease in the crop productivity. High temperatures and skeptical rainfall decrease the grain yield of crops by truncating the length of the growing period. Future projections show that temperature would be incremented by 2.5°C up to 2050. The projected ascend in temperature would cause the high frequent and perpetuate heat waves that can decline the crop engenderment.

The ascension in temperature results in immensely colossal minimization in yield of agronomic crops. Sustaining the crop engenderment under transmuting climate is a key challenge. Ergo, adaptation measures are required to abbreviate the climate susceptibilities.

The deleterious effect of climate change can be mitigated by developing heat-tolerant cultivars and some modification in current engenderment technologies. The development of adaptation strategies in the context of transmuting climate provides the utilizable information for the stakeholders such as researchers, academia, and farmers in mitigating the negative effects of climate change.

8.2. HABITUATING AGRICULTURE TO CLIMATE CHANGE

The vigorous trends in climate change already evident, the likelihood of further changes occurring, and the incrementing scale of potential climate impacts give exigency to addressing agricultural adaptation more coherently. There are many potential adaptation options available for marginal change of subsisting agricultural systems, often variations of subsisting climate risk management.

We show that implementation of these options is liable to have substantial benefits under moderate climate change for some cropping systems. However, there are limits to their efficacy under more astringent climate changes. Hence, more systemic transmutations in resource allocation need to be considered, such as targeted diversification of engenderment systems and livelihoods.

People argue that achieving incremental adaptation action will necessitate integration of climate change-cognate issues with other risk factors, such as climate variability and market peril, and with other policy domains, such as sustainable development.

Dealing with the many barriers to efficacious adaptation will require a comprehensive and dynamic policy approach covering a range of scales and issues, for example, from the construction by farmers of vicissitude in risk profiles to the establishment of efficient markets that facilitate replication strategies. Science, additionally, has to acclimate.

Multidisciplinary quandaries require multidisciplinary solutions, i.e., a fixate on integrated rather than disciplinary science and an invigorating of the interface with decision-makers. A crucial component of this approach is the implementation of adaptation assessment frameworks that are pertinent, robust, and facilely operated by all stakeholders, practitioners, policymakers, and scientists.

Agriculture is the major land use across the globe. Currently, 1.2–1.5 billion hectares are under crops, with another 3.5 billion hectares being grazed. Another 4 billion hectares of forest are used by humans to differing degrees, whereas, away from land, ecumenical fisheries are used very intensively, often beyond capacity.

In order to meet projected magnification in human population and per capita aliment demand, historical increases in agricultural engenderment will have to perpetuate, ineluctably doubling current engenderment. Agriculture is additionally a major economic, gregarious, and cultural activity, and it provides a wide range of ecosystem accommodations.

Importantly, agriculture in its many different forms and locations remains highly sensitive to climate variations, the ascendant source of the overall interannual variability of engenderment in many regions and a perpetuating source of disruption to ecosystem accommodations.

For example, the El Niño Southern Oscillation phenomenon, with its associated cycles of droughts and flooding events, expounds between 15%

and 35% of ecumenical yield variation in wheat, oilseeds, and coarse grains. This subsisting sensitivity expounds why a transmuting climate will have subsequent impacts on agriculture.

Figure 8.1. Crop adaptation to climate change.

Source: Image by unsplash.com.

Hence, it has become critical to identify and evaluate options for habituating to climate transmutation in coming decades. Here we use the term "adaptation" to include the actions of adjusting practices, processes, and capital in replication to the authenticity or threat of climate change, as well as replications in the decision environment, such as transmutations in gregarious and institutional structures or altered technical options that can affect the potential or capacity for these actions to be realized.

People argue there is a vigorous rationale for an incrementing fixate on the adaptation of agriculture to climate change. This need arises from several considerations:

1. Past emissions of greenhouse gases have already committed the globe to further warming of ≈0.1°C per decennium for several decenniums, making some level of impact, and compulsory adaptation replications, already ineluctable.
2. The emissions of the major greenhouse gases are perpetuating to increment, with the resultant transmutations in atmospheric CO_2 concentration, ecumenical temperature, and sea level visually examined today already at the high cessation of those implicatively insinuated by the scenarios considered by the Intergovernmental Panel on Climate Change (IPCC). Furthermore, some climate change impacts are transpiring more expeditious than aforetime considered likely. If these trends perpetuate, then more proactive and expeditious adaptation will be needed.

3. There is currently a lack of progress in developing ecumenical emission-minimization acquiescent beyond the Kyoto Protocol, leading to concerns about the caliber of future emissions and hence climate changes and associated impacts.
4. The high terminus of the scenario range for climate change has incremented over time, and these potentially higher ecumenical temperatures may have nonlinear and increasingly negative impacts on subsisting agricultural activities.
5. Climate changes may additionally provide opportunities for agricultural investment, rewarding early action taken to capitalize on these options.

   There is an immense diversity of agricultural practices because of the range of climate and other environmental variables; cultural, institutional, and economic factors; and their interactions. This designates there is a correspondingly immensely colossal array of possible adaptation options.

The objectives of this topic of the chapter are first to outline these options for cropping and livestock systems, forestry, and fisheries, using the literature on crop yields as an example to assess the benefits of adaptation; and second, to suggest some general pathways that can avail peregrinate from technical assessment of adaptation options to more practical action.

Accordingly, it has been observed that some preconditions for more efficacious uptake of adaptations; develop an adaptation framework to engage all decision-makers (farmers, agribusiness, and policymakers) that builds on the subsisting substantial cognizance of agricultural systems; and outline how science itself needs to adapt to remain relevant in this issue.

The purport of undertaking agricultural adaptation is to efficaciously manage potential climate risks over the coming decades as climate changes. Adaptation research undertaken now can avail apprise decisions by farmers, agrobusiness, and policymakers with implicative insinuations over a range of timeframes from short-term tactical to long-term strategic.

However, it is categorically consequential to align the scales (spatial, temporal, and sectoral) and reliability of the information with the scale and nature of the decision. For example, short-term climate adaptation by farmers may be accomplished by taking into account local climate trends if there is a vigorous correspondence between these trends and projected climate changes, or it may be via climate forecasting at scales from daily to interannual.

Nevertheless, some farmers may find inhibited utility in long-term projections of climate, given the high uncertainties at the finer spatial and temporal scales at which their decisions are made. In contrast, the general trends at more astronomically immense time and spatial scales able to be more reliably projected with current climate models may be quite serviceable for input into policy and investment analyses, provided potentially critical factors are incorporated such as transmutations in climate extremes.

A paramount benefit from adaptation research may be to understand how short-term replication strategies may link to long-term options to ascertain that, at a minimum, management and/or policy decisions implemented over the next one to three decennials do not undermine the competency to cope with potentially more astronomically immense impacts later in the century. In the sections below, we endeavor to identify other key benefits from an incremented fixate on climate change adaptation.

8.2.1. Keeping Policy Germane

At the current relatively early stage of the debate, it is understandable that climate change adaptation is largely being dealt with in isolation from other issues. However, over time, this situation needs to evolve so that climate change is linked with a much broader set of policies.

In particular, there is a desideratum for linkage with subsisting policies on climate risk such as those on drought or structural adjustment, which otherwise may become poorly targeted. Climate change will require these policies to become more dynamic, to cope with the high caliber of dubiousness in the timing and magnitude of potential climate changes and the expeditiously evolving erudition base.

In addition to this, climate change adaptation policies will interact with, depend on, or perhaps even be just a subset of policies on sustainable development and natural resource management, such as those indispensable to regulate genetically modified organisms, bulwark human and animal health, and foster governance and political rights, among many others.

This process is often referred to as the "mainstreaming" of climate change adaptation into policies intended to enhance broad resilience to jeopardize or to promote sustainable development. The critical issues of how climate change and adaptation may affect aliment security and trade and the peril of malnourishment are dealt with in a companion paper.

8.2.2. Apprising Mitigation Targets

Importantly, identifying and evaluating possible adaptation strategies are of fundamental value to determine a set of dynamic climate policy options that lead to the "avoidance of perilous anthropogenic interference" component of the Coalesced Nations Framework Convention on Climate Change.

This is because maximizing societal welfare under future climate risk will likely involve an amalgamation of both mitigation and adaptation; the percentage contribution of each strategy will depend on monetary and nonmonetary cost/benefit analyses.

For example, we would expect the size and cost of the adaptation task to be lower if there is efficacious, but perhaps costly, mitigation and higher if there is no mitigation. Similarly, the benefits of adaptation will be a function of the nature of climate change and the scale of impact.

Consequently, inadequate consideration of adaptation options could result in the susceptibility to climate change being significantly overstated, giving ascend to more astringent mitigation targets. Supplementally, mitigation policies can affect the range of adaptation options that practitioners have at their disposal (e.g., subsidizing biofuel engenderment vigorously influences the market for agricultural engender).

Another perspective is that implementing efficacious adaptation can "buy time" until an efficacious mitigation replication can be mounted. Hence adaptation analyses may be habituated to appraise both the magnitude and timing of mitigation. Achievement of this intricate task of efficaciously integrating mitigation impacts and adaptation to apprise public policy development remains a paramount challenge for the scientific community, albeit some studies are now emerging.

This interaction of science and policy needs to evolve as the scientific cognizance base changes and may adscititious focus attention on the paramountcy of integrative rather than disciplinary science within the science-policy interface.

8.2.3. Apprising Investment

Adaptation analyses can also avail assistance to the authorities and industry of the investment or disinvestment decisions they require to make now or in the near future in cognation to climate-sensitive aspects of their portfolios.

In particular, this applies to long-term investments such as plant and animal breeding programs; building capacity in the scientific and user communities; developing quarantine systems; establishing perennial crops and forest plantations; purchasing or selling land; or building (or decommissioning) major infrastructure such as dams and dihydrogen monoxide distribution systems, flood mitigation works, and storage and convey facilities.

Climate risks are, of course, only one consideration within more involute decision-making processes. For example, in Western Australia, incremental risk of drought under ecumenical warming was integrated with projections of population magnification, economic development, and convivial norms in cognation to dihydrogen monoxide use, resulting in the construction of a major incipient dam and development of other incipient dihydrogen monoxide sources.

8.2.4. Rewarding Early Adopters

Participatory research into climate change adaptation options can avail agricultural decision-makers realize that acting on the subsisting trends in climate now is liable to be to their advantage. For example, in north-east Australia, crop management that has perpetually adjusted to the progressive abbreviation in frost risk experienced over the past several decades can virtually double gross margins when compared with management predicated on either the long-term risk or management that does not consider frost peril.

Participatory engagement with decision-makers, by bringing their practical erudition into the assessment, can withal identify a more comprehensive range of adaptations than are typically explored by scientists, as well as being able to assess the practicality of options and contribute to the more authentic assessment of the costs and benefits involved in management or policy change.

8.2.5. Fixating on Climate Risk Management

Determinately, it should be perceived that "adaptation" is a perpetual process that is a component of good risk management, whereby drivers of peril are identified, and their likely impacts on systems under alternative management are assessed. In this reverence, adaptation to climate change is akin to adaptation to climate variability, vicissitudes in market forces (cost/price ratios, consumer demands, etc.), or institutional or other factors.

Differences may be in the rate of realized climate change, compared with how expeditious we are able to implement needed solutions. Isolating climate change from other drivers of peril may be subsidiary, especially during the initial stages of assessment when vigilance of the relative paramountcy of this jeopardy factor is still low.

Operationally, however, translating adaptation options into adaptation actions requires consideration of a more comprehensive risk management framework. This would sanction exploration of quantified scenarios dealing with all of the key sources of jeopardy, providing more efficacious decision making and learning for farmers, policymakers, and researchers: an incrementation in "climate knowledge."

8.3. CLIMATE CHANGE IMPACTS AND ADAPTATION STRATEGIES FOR AGRONOMIC CROPS

Climate change and variability are the authentic threats to agriculture and pabulum security. Extreme weather events and skeptically in rainfall patterns are negatively affecting the agricultural crops. The evidence of ecumenical trends in rainfall are obscure due to immensely colossal regional gaps in spatial coverage and temporal shortfalls in the data. Owing to these vicissitudes, the drought is more prevalent in many regions of the world including Pakistan.

Finding evidence reported that high temperature and uneven distribution of rainfall have negative effects on crop productivity all over the world. These transmutations in weather and climate are liable to affect the victuals security of the developing world where an astronomically immense fraction of ever-incrementing population is already fronting hunger, insecure and insalubrious pabulum.

Warming of weather and climate systems can result in highly corresponding transmutations in the occurrence of extreme events, including ascend in temperature, uneven rainfall patterns. These extreme events occur due to shifts in their distribution, or to transmute in the shape of distribution. Sundry studies suggest that a shift in mean accounts for much of the transmutation in visually examined temperature extremes.

Comparisons of sundry studies showed that both daily maximum (T_{max}) and minimum (T_{min}) temperatures have shifted toward higher values in all regions. These shifts in temperatures and rainfall significantly affected the cropping patterns, crop yields, and phenology.

The Intergovernmental Panel for Climate Change (IPCC) has found evidence of expedited ecumenical warming, climate variability and transmute since the early 1990 s. The Intergovernmental Panel for Climate Change reported that average ecumenical temperature in the last 100–150 years has incremented by 0.76°C.

Figure 8.2. Climate change impacts and adaptation strategies for agronomic crops.

Source: Image by unsplash.com.

The variability in temperature altered the phenology of crop, i.e., leaf development, anthesis, harvest, fruit engendered and in asynchrony between anthesis and pollinators. The variable temperature range resulted in high respiration rates, abbreviation in pollen germination, shorter grain filling period, lesser biomass engenderment and low yields.

High temperature above 35°C in amalgamation with high sultriness and low wind speed caused a 4°C increase in temperature, resulting in floret sterility in cereals and fruits. Climate change impact assessment provides the scientific substratum for the development of adaptations to offset the negative impacts of climate change. Keeping in view, the current study was orchestrated to assess the impacts of climate change and adaptations strategies for agronomics crops.

8.4. IMPACT OF CLIMATE CHANGE ON CROP ENGENDERMENT

Ecumenical atmospheric concentrations of greenhouse gases have significantly incremented relative to preindustrial time. As a result, greenhouse gas forcing is the main cause of the warming of the atmosphere during the past decades.

This warming is expected to substantially alter the climate system and transmute ecumenical pabulum engenderment, mainly because temperatures are prognosticating to increment, which in turn will alter the precipitation pattern and increment the frequency of extreme events such as drought. Man-made greenhouse gas emissions as a result of industrialization and urbanization have made paramount contributions to ecumenical warming and further transmutations in the ecumenical climate.

As a result, ecumenical temperature rose by 0.83°C from 1906 to 2010. Ecumenical warming withal causes transmutations in precipitation levels and patterns due to higher evapotranspiration and dihydrogen monoxide vapor amounts in the atmosphere with several implicative insinuations for the ecumenical hydrological cycle.

As the major dihydrogen monoxide consumer of the developing world and some developed countries, agriculture is one of the most vulnerable susceptible dihydrogen monoxide sectors to climate change. Dramatic population magnification, associated with truncation of productive land area and dihydrogen monoxide resources, exerts extra pressure on the agricultural sector. To ascertain the sustainability of agriculture, studying the possible climate change impacts on this sector is essential.

Rate of plant magnification and development is dependent upon the temperature circumventing the plant, and each species has a concrete temperature range represented by a minimum, maximum, and optimum. The expected transmutations in temperature over the next 30–50 years are predicted to be in the range of 2–3°C Intergovernmental Panel for Climate Change.

Figure 8.3. Impact of climate change on crop engenderment.

Source: Image by unsplash.com.

Heatwaves or extreme temperature events are projected to become more profound, more frequent, and last longer than what is being currently been optically canvassed in recent years. Extreme temperature events may have short-term durations of a few days with temperature increases of over 5°C above the mundane temperatures. Extreme events occurring during the summer period would have the most dramatic impact on plant productivity.

A recent review by a well-known researcher named as, Barlow, which stated that the effect of temperature extremes, frost and heat, in wheat (*Triticum aestivum* L.) revealed that frost caused sterility and abortion of composed grains while extravagant heat caused minimization in grain number and truncated duration of the grain-filling period.

Analysis revealed that daily minimum temperatures will increment more expeditiously than daily maximum temperatures leading to the incrementation in the circadian mean temperatures and a more preponderant likelihood of extreme events and these vicissitudes could have detrimental effects on grain yield.

If these vicissitudes in temperature are expected to occur over the next 30 years, then understanding the potential impacts on plant magnification and development will avail develop adaptation strategies to offset these impacts.

8.5. ADAPTATION STRATEGIES FOR AGRONOMIC CROPS

Climate change adaptation is the action to ecumenical warming, which avails to minimize the susceptibilities in the convivial and biological system. The main objective of adaptation strategy is to build the resilient in societies against climate change.

Agriculture sector is highly vulnerable susceptible to transmuting climate. Extreme weather conditions and transmuting patterns of precipitation affects the crop development, magnification, and yield of crops. High temperature at critical magnification stages could abbreviate the grain filling duration causing the grains sterility and consequently yields abbreviations.

In order to evade the perils in agriculture associated with climate change (CC), adaptation is the key factor that could be used to mitigate the negative of climate change. Adaptation strategies provide an opportunity to address the climate change challenges and to sustain the crop engenderment. In the recent year, climate change adaptation has been explored by the farmers in many ways. For example, in Pakistan and Brazil farmers have habituated the climate change variability by adjustment of planting time and optimization of plant populations. Adjustment of planting date is paramount to explore the plenary potential of the crop. High temperature at the grain filling stage, abbreviates the time for grain filling that leads to the decrement of the yield.

Figure 8.4. Adaptation strategies for agronomic crops.

Source: Image by unsplash.com.

Adjusting the planting time with the onset of rains and heat waves would decrement the yield losses. A number of plants per unit area play a vital role for higher yield in crops, especially wheat. The number of productive tillers dies or remains unproductive due to variation in temperature and moisture stress. The optimum plant population compensates the yield loss.

The development of ameliorated varieties such as early maturing, drought and heat tolerant are indispensable to sustain the productivity under transmuting climate. The incipient cultivars would increment the engenderment per unit area under moisture stress and extreme temperatures.

Methane gas is engendered from the flooded rice. Flood dihydrogen monoxide in rice blocks the oxygen to perforate in soil that engenders the propitious condition for bacteria that emit the methane gas. So incipient methods of planting like direct-seeded rice and system of rice intensification with Alternate wetting and drying abbreviate the methane emission and increment the dihydrogen monoxide use efficiency.

Precision management of nutrients can increment the resilience in the crops by incrementing the efficiency of fertilizers. Precision management of fertilizers in crops, especially maize, truncated the utilization of fertilizers that would enhance the engenderment and soil health that lead to decrement the emission of greenhouse gases (GHGs).

Ratoon crop of sugarcane is more adaptive to climatic susceptibilities. Fuel consumption is less for tillage practices, and less soil is perturbed that lead to abbreviate the GHGs emission. Pit planting is an incipient evolutionary method in sugarcane. In this method, sugarcane seedlings are grown in a minute pit under field conditions.

This method ameliorates the aeration and solar radiation that lead to increment the quality of cane juice and number of canes for milling. Weeds are an issue in chickpea cultivation. Weeds compete with the chickpea plants for dihydrogen monoxide and nutrients that minimize the magnification and yield of chickpea. So integrated weed control ameliorates the yield. GHGs emissions are additionally truncated due to less utilization of synthetic weedicides.

8.6. PLANT YIELD AND CLIMATE CHANGE

Plant physiology has been greatly influenced by climate variability by several denotes. Environmental extremes and climate variability enhanced the chances of numerous accentuates on plants. Climate change affects crop engendered by betokens of direct, indirect, and socioeconomic effects. Furthermore, climate change (drought, flood, high temperature, storm, etc.) events are incremented dramatically as reported by Pabulum and Agriculture Organization (FAO).

Boyer reported that the climate changes have abbreviated the crop yield up to 70% since 1982. According to the study of FAO 2007, all cultivated areas in the world are affected by climatic changes, and only 3.5% of areas are safe from environmental circumscriptions.

Whereas the outcomes of abiotic accentuates on crop yield are hard to calculate accurately, it is believed that abiotic stresses have a substantial influence on crop engenderment depending upon the extent of damage to the total area under cultivation. In the future, the productivity of the major crops is estimated to drop in many countries of the world due to ecumenical warming, dihydrogen monoxide shortage, and other environmental impacts.

Predicated on national crop yields and questionnaire surveys, immensely colossal differences in susceptibilities to current climate changes were detected across Europe. In Northern Europe, the short duration for crop development and cool temperature are the major concerns, while the temperature extremes and low rainfall limits the crop productivity in Southern Europe, albeit the most negative effects will be found for the continental climate in the Pannonian zone, which includes Hungary, Serbia, Bulgaria, and Romania.

It was said that the enhancement of greenhouse gas emissions and abrupt climatic changes will occur that may increase the crop yield in North-Western Europe and decrement the crop yield in the Mediterranean area. Wheat engenderment is heavily affected by the temperature extremes due to climate vicissitude in many countries, and may minimize the crop yield by 6% for each °C ascend in temperature.

Drought and high temperatures are key stress factors with high impact on cereal yields, and Rubisco, the central enzyme of photosynthesis, is disrupted if the temperature increases from 35°C, and ceases the photosynthetic process.

Figure 8.5. Plant yield and climate change.

Source: Image by unsplash.com.

Gong et al. (1997) reported the negative influence of heat accentuate on antioxidant enzymes in *Zea mays*. The amalgamated impact of heat and drought accentuates on crop yield have been examined in sorghum, maize, and barley. It was revealed that the accumulated effect of heat and drought stress had more damaging outcomes as compared to individual stress. Xu and Zhou (2006) subjected the Leymus Chinensis under the cumulated stresses of drought and heat and found that the function of Photosystem II (PSII) decremented.

Due to climate change, dihydrogen monoxide deficit and temperature extremes influence the reproductive phase of plant magnification. It was described that the flower initiation and inflorescence is lamentably affected by the dihydrogen monoxide stress in cereals.

Similarly, if the temperature increases about 30°C during floret development it can cause sterility in cereals. During the meiotic phase, wheat and rice suffered from the 35–75% minimization in grain set due to dihydrogen monoxide deficit. In rice, drought stress greatly perturbs the process of fertilization and anthesis.

Due to dihydrogen monoxide deficit, the harvest index is truncated to 60% and decreases the grain set. The cocoa yield has been significantly truncated by the major drought spells in West Africa during the 1980 s El Niño years. It has been estimated that agricultural engendered could abbreviate to 25.7% by 2080 due to climate change and maize will be the most affected crop in Mexico.

A study predicated on ECHAM6 climate data was analyzed for North German Plains during two different time durations: 1981–2010 and 2041–2070. The results showed that if the yield for winter wheat is to be sustained, dihydrogen monoxide availability must be ensured.

Zhao et al. (2017) carried an experiment to analyze the climate change impact on major crop yields and showed considerable yield reductions of 6%, 3.2%, 3.1%, and 7.4% in wheat, rice, soybean, and maize, respectively. To tackle the climate change, incipient revelations in genomics are enabling climate-keenly intelligent agriculture by developing climate-resilient crops.

Drought stress influences wheat during all developmental stages, but grain formation and the reproductive stage are the most critical ones. Wheat yield was decreased from 1% to 30% during the mild drought stress at postanthesis while this truncation increased up to 92% in case of perpetuated mild drought stress at flowering and grain formation.

Drought stress has greatly minimized the yield of consequential grain legumes. Mashbean (*Vigna mungo* L.) yield has been abbreviated by drought stress from 31% to 57% during the flowering stage, while a 26% abbreviation was reported by drought stress during the reproductive phase. Maleki et al. (2013) reported that the soybean yield has been largely affected by drought stress and a 42% minimization was visually examined during the grain filling stage of soybean.

Schlenker and Roberts (2009) described that maize yield was incremented at an optimum temperature of 29°C but a further increase in temperature hampered the yield of maize. Every 1°C ascend in temperature was found to negatively influence the maize yield. Similarly, it was reported that yield in maize decremented by 8.3% with every 1°C ascend in temperature from the optimum magnification temperature. Brown (2009) reported that wheat yield decremented by 10% with every 1°C increase in temperature.

In another report, it was revealed that a 3–4% minimization in wheat yield takes place for every 1°C increase in temperature. Easterling et al. (2007) described that a 2°C increase in temperature caused 7% truncation in yield while a further increase in temperature to 4°C decreased the yield by up to 34% in wheat. Similarly, rice yield decreased by 2.6% for every 1°C rise in temperature.

In sorghum, yield was minimized by 7.8% due to a 1°C increase in temperature. In sorghum, dihydrogen monoxide shortage is another immensely colossal issue reported in most of the world's top engenderer countries. Schlenker and Roberts (2009) revealed that the threshold temperature for soybean is 30°C; an ascension in temperature to the optimum level incremented soybean yield but after that caliber, further ascend in temperature abbreviated the yield abruptly.

Eastburn et al. (2010) reported that the ascension in ozone and the concentration of the carbon dioxide (CO_2) in the atmosphere influenced the disease type, and with a perpetual ascend in temperature, disease susceptibility in soybean was enhanced.

This ascending concern was revealed in the growing quantity of research papers fixated on abiotic quandaries after the crucial review by Kitano on systems biology. The magnitude of research studies has incremented dramatically cognate to biotic and abiotic stresses in plants by applying different strategies.

8.7. HOW WILL CLIMATE CHANGE AFFECT CROPPING SYSTEMS?

Beyond its direct effects on weather, climate change will increment both abiotic stresses, such as drought, and biotic stresses, such as pest and crop disease pressures, on agricultural systems. Of greatest concern and largely unknown, are the influences that interactions among variants of stresses will have on crops.

Drought is expected to inhibit the productivity of over half the earth's arable land in the next 50 years (Cattivelli et al., 2008; Sinclair, 2010), and competition for dihydrogen monoxide between urban and agricultural areas will compound issues of dihydrogen monoxide availability (Rosegrant et al., 2009).

While the utilization of brackish and saline dihydrogen monoxide could additionally alleviate the world's dihydrogen monoxide quandaries, this option is only possible with the development of salt-tolerant crops or management practices that alleviate salt stress.

As a result, to constrain the impact of drought, there is a clamant desideratum for crop varieties and cropping systems that conserve dihydrogen monoxide and retain yield during periods of dihydrogen monoxide scarcity.

Developing these crops is arduous, however, because of the interplay of crop replication systems to drought at the genomic, metabolic, biochemical, and physiological levels.

To make drought-tolerant varieties available to farmers, interdisciplinary teams of scientists working at the cellular, plant, and field scales must collaborate to discover ways to manipulate these intricate, multilevel processes and amend crop replication.

8.7.1. Temperature

Temperature influences the magnification and development of all crops, shaping potential yield throughout the growing season. Current temperatures in the Midwest of the U.S. are optimum for engenderment, while the Southern U.S. temperatures already exceed the optimum.

Temperature events higher than mundane are expected to truncate cereal and grain legume yields (Hatfield et al., 2011). Ascended temperatures are kenned to abbreviate the grain-filling period, for example, and to abbreviate pollen viability and weight gain in grain (Boote and Sinclair, 2006; Hatfield et al., 2011).

Moreover, temperature changes can result in warmer, less astringent winters, which sometimes sanction diseases and pests to survive and overwinter, incrementing the likelihood of minimized yield during the next cropping season. For all these reasons, habituating crops and cropping systems to incipient seasons and temperatures will require region-concrete crop adaptation strategies.

8.7.2. Carbon Dioxide

Carbondioxide(CO_2)isfundamentaltocropcarbohydrateengendered(paramount for crop productivity and yield) and overall plant metabolism. It additionally plays a paramount role in climate change. Atmospheric CO_2 concentrations have ascended dramatically over the past 200 years and may reach 450–1,000 μmol by the terminus of this century, according to the Intergovernmental Panel on Climate Change (IPCC, 2007). Ascending CO_2 levels will likely boost the overall productivity of many crops; albeit paramount tropical grasses like maize, sugarcane, and sorghum, and some cellulosic biofuels crops, don't respond as well to ascended CO_2 levels (visually perceive sidebar on C3 and C4 crops). Increases in productivity could be offset, though, by pressures such as insect and fungal pests, ozone, and more variable precipitation, albeit the degree to which this occurs will depend on the physiology and biochemistry of each crop (Ziska, 2008; Taub, 2010). Crop adaptation will be critical to ascertain that crops can maintain, or even increase, productivity amidst a CO_2-enriched environment.

8.7.3. Ozone

Ozone (O_3)—a consequential greenhouse gas and agricultural pollutant—continues to increment because of fossil fuel combustion (Staehelin et al., 2001; Krupa et al., 2000). While levels of CO_2 will ascend rather uniformly around the globe, O_3 concentrations will vary regionally and subsist to a more preponderant extent around industrialized areas (Jaggard et al., 2010).

Crops take ozone into their leaves during photosynthesis, where the gas lowers photosynthetic rates and expedites leaf death, affecting crop maturity and productivity (Krupa et al., 2000). The rate at which crops take up O_3 depends on the O_3 concentration in the air as well as the aperture and closing of the stomata of leaf pores. At a present interval of time, ecumenical yield losses due to ozone are estimated at approximately 10% for wheat and soybean, and 3–5% for rice and maize (Van Dingenen et al., 2009). Modern implements of crop breeding will avail to apprise scientists about different approaches for developing crops that can thrive despite exposure to incrementing concentrations of O_3.

8.7.4. Biological Stresses

Biological accentuates on cropping systems include weeds, insects, viruses, bacteria, and fungi. Temperature is considered the most consequential factor in determining how insects affect crop engenderment and yield (Coakley et al., 1999). For example, some populations of insect species, such as flea beetles, are exhibiting denotements of overwintering because of warmer winter temperatures (Harrington et al., 2001; Wolfe et al., 2007).

Viral, bacterial, and fungal pathogens additionally respond greatly to temperature, as well as sultriness and rainfall. Thus, as the growing season lengthens and winters moderate due to climate change, pressures from plant, microbial, and insect pests are expected to ascend due to an incremented capacity for overwintering, more preponderant kineticism of organisms, and expanded adaptation zones.

8.8. HOW CAN PEOPLE ACCLIMATE CROPS AND CROPPING SYSTEMS TO CLIMATE CHANGE?

The climate has always been in a state of flux, but the current rate of change is much more expeditious, and the range of weather variables much broader than ever visually perceived in modern agriculture. Today, two primary approaches subsist for habituating crops to these conditions:

1. ameliorating subsisting crop cultivars and developing incipient crops, and
2. devising incipient cropping systems and methods for managing crops in the field.

These approaches include the categorical strategies discussed below.

8.8.1. Strategies for Ameliorating Subsisting Cultivars and Developing Incipient Crops

Develop incipient crops. Incipient crops will likely play a key role in maintaining and incrementing agricultural engenderment. Domestication commenced only 5,000 to 12,000 years ago for our oldest crops such as maize, wheat, potatoes, and sorghum, while blueberries and wild rice were domesticated more recently (Harlan, 1991).

Today, some scientists are crossing wild, perennial relatives of crops such as maize, millet, rice, sorghum, sunflower, and wheat with their annual, domesticated counterparts, to develop perennial grain crops (Cox et al., 2006). Adscititious, a growing interest in bioenergy has supplementally enheartened the domestication and breeding of C4 grasses, including switchgrass and Miscanthus (Bransby et al., 2010). Domestication and breeding of incipient crops is a long-term solution, requiring many years of effort before formal testing can be performed.

Integrate salutary traits into subsisting crops through utilization of germplasm amassments, cognate datasets, and breeding. Historically, crop scientists have identified and selectively habituated crops to exhibit desirable traits that sanction crops to achieve optimum yields while withstanding stresses, such as drought, heat, and waterlogging.

The prosperity and haste of breeding efforts depend, however, on the competency of breeders to access optimal germplasm and quality information about germplasm material. Today's breeders rely on genetic and environmental information in both private and public germplasm accumulations, such as the U.S. Department of Agriculture (USDA) Agricultural Research Service's National Plant Germplasm System.

In order to fortify perpetual amelioration of germplasm that can be habituated to develop cultivars acclimated to climate change, there is a desideratum to acquire, preserve, evaluate, document, and distribute plant genetic resources for a wide range of crops and their wild relatives.

Supplementally, well-preserved information can sanction scientists to employ modern biotechnology methods to screen crop traits – these advances are already transmuting how germplasm banks are used. Expanded utilization of these resources and methods will avail researchers more expeditiously identify adaptive traits, represented by genes or groups of genes, which contribute to stress resistance.

Use incipient technologies – image-predicated quantifications, high-throughput DNA sequencing, databases, and statistical models. These technologies are transmuting the way that germplasm banks are used and surfacing incipient ways to use information stored in germplasm banks to ameliorate crop performance. For today's challenges, it is essential to have public access to precise information, sanctioning for identification of adaptive traits represented by genes or groups of genes that contribute to abiotic or biotic stress resistance. Utilizing crop breeding alone or in partnership with incipient biotechnology, scientists can move expeditiously identify cultivars with enhanced productivity even in the presence of drought, heat, dihydrogen monoxide logging, and frost pressures relative to the current cultivars available. Identify crop germplasm that abides drought, heat, and dihydrogen monoxide-logging. Yield drops when crops experience drought, extravagant heat, or surplus dihydrogen monoxide deviating from the optimum for magnification during key stages, including pollination, flowering, and filling periods, when carbohydrates and nutrients assimilate inside grain, tubers, or fruit.

Multiple molecular markers can be statistically associated with some of these traits to sanction cull to be performed without testing in the breeding environment. Cultivars are now being developed which are tolerant to excess heat during pollination for cowpea and corn, and flooding early in the growing season for soybean and rice (Hall, 2004; VanToai et al., 2010; Bailey-Serres et al., 2010).

Maize hybrids are now being developed that have a better synchronization of pollination and flowering under heat and dihydrogen monoxide stress (Ribaut et al., 2009). Cultivar differences for heat tolerance subsist in some crops such as rice, cowpea, and peanut, but erudition about the effects of prodigiously high temperatures is very inhibited because diverse germplasm has not been extensively screened. Perpetuated work in this area will provide incipient germplasm for plant breeders to incorporate into productive varieties for farmers.

Expand field-level evaluations of crop germplasm. The modern crop breeder's toolkit enables consideration of the entire world's genetic resources for use in crop genetic amendment. Incipient technology in cumulation with immensely colossal scale field phenotypic evaluations will likely avail disclosure of antecedently unknown genetic sources and genomic regions on DNA associated with abiotic stress tolerance (Collins et al., 2008).

These genetic resources and erudition will foster consequential leaps in abiotic stress tolerance. This, in turn, will avail scientists discern molecular to physiological mechanisms underpinning drought tolerance that are now poorly understood because agriculturally paramount stress tolerance is a relatively incipient area of science (Sinclair, 2011).

Surmounting this 'knowledge gap' will unlock the facility of both applied and rudimentary research groups to develop long-term strategies that can maximize the distribution of incipient, amended cultivars. Consequently, expanded field-predicted phenotypic research programs and cognate breeding efforts must be approximately coordinated with the full spectrum of crop development sciences, engaging breeders, physiologists, and geneticists.

8.8.2. Climate-Smart Crop Production Practices and Technologies

This topic of the chapter presents management practices and technologies for climate change adaptation and mitigation. It covers practices with an explicit fixate on adaptation to categorical climatic stressors, and practices that simultaneously abbreviate engenderment risks and lower greenhouse gas emissions. Most of these practices obviate soil damage that relinquishes carbon and dihydrogen monoxide into the atmosphere; promote soil and dihydrogen monoxide conservation; and increase productivity.

8.8.3. Utilization of Quality Seeds and Planting Materials of Well-Acclimated Crops and Varieties

An indispensable input for climate-keenly intellective crop engenderment is quality seeds and planting materials of well-habituated varieties. It is infeasible to harvest good crops with deplorable seeds (FAO, 2011).

National, regional and international plant breeding efforts customarily involve multilocational tribulations and seek to develop crop varieties that are resistant to climate-cognate phenomena and more efficient in their utilization of resources to truncate their impact on the agricultural ecosystem and the wider environment.

Resistance to drought, salinity and flooding are the most prevalent climate-cognate traits for which crop varieties are bred. Other more location-concrete factors include higher frequencies of frosts at the seedling and/or pollination stages; high temperatures at the grain-filling stage; heftily ponderous rains that compress the soil; and alternate light rains and sultry temperatures that stimulate seed germination but avert the establishment of seedlings.

The development, official release, and registration of well-acclimated crop varieties are the steps taken toward the ultimate goal of ascertaining farmers have access to quality seeds and planting materials. However, achieving this ultimate goal additionally requires a reliable mechanism for distributing the seeds of the most opportune varieties to farmers. Farmers obtain seeds from formal systems and/or informal systems.

Formal seed systems are organized and underpinned by statutory requisites that ascertain the seeds that farmers use pass through standardized quality assurance mechanisms. Distributing the amended varieties to farmers through formal systems is relatively straightforward. But, farmers, especially in developing countries, additionally obtain seeds from multiple unregulated sources.

These sources can include preserving seeds from their own harvests, purchasing seeds from local markets, and exchanging seeds with family members and neighbors. This way of obtaining seeds constitutes the informal system or farmers' seed system. Informal systems predominate in developing countries where crop engenderment systems are the most vulnerable susceptible to extreme weather events.

For such engenderment systems, which are typically characterized by low-input agriculture, diminutive-scale holdings and constrained market engagement, it is concretely consequential to fortify community-predicated seed engendered and distribution channels. In components of the world where climate change is expected to have the greatest impact, most of the seeds sourced through community-based distribution systems are paramount pabulum security crops.

These crops include beans, peanuts, cassava, cowpeas, open-pollinated maize, saccharine potato and yams. Minuscule- and medium-scale enterprises are efficacious designates for ascertaining that quality seeds of the most felicitous varieties are available to minuscule-scale farmers and are within facile reach in their communities.

Ideally, irrespective of the distribution system, seeds that are designated to be sold or otherwise distributed for planting must have their quality assured. There are different options for achieving quality assurance. In the formal system, there is official control and inspection, customarily by regime agencies or accredited entities such as farmers' sodalities and seed companies, which result in certified seeds.

A less authoritatively mandating mechanism is the Quality Declared Seeds and Quality Declared Planting Materials systems, developed under the auspices of FAO. These options ascertain that costs associated with standard certification processes do not obstruct the availability of quality seeds.

8.9. SEED SYSTEMS

A seed system encompasses all the stakeholders (individuals, organizations, and institutions) that are involved in the development and dissemination of crop varieties; the production, multiplication, processing, storage, distribution and marketing of seeds and related practices and processes; and the prevailing policies, regulations and laws. The following components of a seed delivery system are critical in formulating for climate-smart agriculture strategies.

8.9.1. Conservation of Plant Genetic Resources for Food and Agriculture

Conserved and characterized plant genetic resources for food and agriculture need to be available for use as 'raw materials' for the development of varieties that are resistant to abiotic and biotic stresses. In order to address the challenges posed by climate change, there is the increasingly urgent need for the investment of greater resources and efforts in safeguarding the widest possible diversity of plant genetic resources for food and agriculture in their natural habitats in situ, on farms and in gene banks ex-situ.

The diversity of crop wild relatives, an important source of heritable traits for crop improvement, could be eroded as their natural habitats are lost due to climate change.

Figure 8.6. Seed system.

Source: Image by unsplash.com.

8.9.2. Crop Varietal Development

Plant breeders must develop an increasingly diverse portfolio of varieties of an extensive range of crops in order to acclimate engenderment systems to climate change. Engendering novel varieties will most often depend on obtaining heritable variations, especially from the nonacclimated materials, including crop wild relatives, that are not customarily used by breeders.

This will involve institutionalizing and ameliorating capacities for pre-breeding activities in which germplasm curators and breeders collaborate to identify the carriers of desirable traits, evaluate these putative parents and cross promising ones with elite lines to engender intermediate breeding materials.

It may be obligatory to engender novel variations that are absent in the gene pool through induced mutations and the application of biotechnological procedures, such as genetic engineering and genome editing. High-throughput genotyping and phenotyping platforms are being used more and more to make the processes for developing crop varieties, including prebreeding, more efficient.

In farmer participatory plant breeding, farmers and plant breeders collaborate in crop varietal development. Because the perspectives of the farmers contribute to the decisions about which varieties are proposed for official release and registration, participatory plant breeding is an efficacious way to achieve demand-driven crop amendments for adaptation to climate change, especially in developing countries.

8.9.3. Seed Production and Delivery

An efficacious agricultural extension system and a responsive seed distribution system are needed to enable farmers to access quality seeds and planting materials of well-acclimated crop varieties at affordable prices and in a timely manner.

1. Farmers are more disposed to use an incipient variety when they have trusted information and are confident the incipient variety will meet their desiderata. The extension system is categorically consequential to engender data about the performance of varieties. For instance, the FAO Farmer Field Schools approach has proven categorically efficacious in using demonstration plots to showcase the advantages of well-acclimated crop varieties to communities of minuscule-scale farmers.
2. A responsive seed distribution system requires national policies, strategies, regulations and licit frameworks that cater to both the informal and formal seed systems and agonize their generally complementary roles
3. Climate change is expected to increase the frequency and intensity of extreme weather events, which will trigger crises that threaten the immediate pabulum security of sizably voluminous populations and possibly spark famines. These events will additionally affect farmers' faculty to obtain quality seeds and planting materials, which will jeopardize the prosperity of subsequent cropping seasons.

8.10. CONCLUSION

In the conclusion of the chapter, this chapter discussed about the various aspects of sudden changes in climate and how it influences the production of crop. This chapter also discussed about the habituating agriculture to climate change, such as keeping policy germane, apprising mitigation targets, apprising investments, rewarding early adopters, and fixating on climate risk management.

This chapter also discussed several impacts of the sudden changes in the climate, and various different adaptation strategies for agronomic crops in order to address these climate changes. Through the end of the chapter, it has been discussed that how will climate change affect cropping systems such as temperature, carbon dioxide, ozone, and other various biological stresses.

This chapter also discussed about how can people acclimate crops and cropping systems to climate change, such as various strategies for ameliorating subsisting cultivars and developing incipient crops, climate-smart crop production practices and technologies, and utilization of quality seeds and planting materials of well-acclimated crops and varieties. This chapter also discussed about the significance of seed systems which includes conservation of plant genetic resources for food and agriculture, crop varietal development, and seed production and delivery.

REFERENCES

1. Ahmed, I., Ullah, A., Rahman, M. U., Ahmad, B., Wajid, S., Ahmad, A., & Ahmed, S., (2019). Climate change impacts and adaptation strategies for agronomic crops. *Climate Change and Agriculture.* [online] Available at: https://www.intechopen.com/books/climate-change-and-agriculture/climate-change-impacts-and-adaptation-strategies-for-agronomic-crops (accessed on 7 July 2021).
2. Baudoin, W., Calles, T., Fredrix, M., Gbehounou, G., Mpagalile, J., Phillips, S., Santos, V. S., et al. (n.d). *B1 Climate-smart Crop Production.* [e-Book] Available at: http://www.fao.org/climate-smart-agriculture-sourcebook/production-resources/module-b1-crops/b1-overview/en/?type=111 (accessed on 7 July 2021).
3. Chloupek, O., & Hrstkova, P., (2005). Adaptation of crops to environment. *Theoretical and Applied Genetics,* [online] *111*(7), 1316–1321. Available at: https://pubmed.ncbi.nlm.nih.gov/16187121/ (accessed on 7 July 2021).
4. Hatfield, J., & Prueger, J., (2015). Temperature extremes: Effect on plant growth and development. *Weather and Climate Extremes,* [online] *10*, 4–10. Available at: https://www.sciencedirect.com/science/article/pii/S2212094715300116 (accessed on 7 July 2021).
5. Howden, S., Soussana, J., Tubiello, F., Chhetri, N., Dunlop, M., & Meinke, H., (2007). Adapting agriculture to climate change. *Proceedings of the National Academy of Sciences,* [online] *104*(50), 19691–19696. Available at: https://www.pnas.org/content/104/50/19691 (accessed on 7 July 2021).
6. Kelleher, F., (n.d). *Chapter 3 Crop Adaptation.* [online] Cdn.csu.edu.au. Available at: https://cdn.csu.edu.au/__data/assets/pdf_file/0006/2805558/Chapter3_Kelleher.pdf (accessed on 7 July 2021).
7. Kenneth, J. B., Amir, M. H. I., Renee, L., Rebecca, L. M., Charlie, M., Seth, C. M., James, E. S., et al. (n.d). *Position Statement on Crop Adaptation to Climate Change.* [e-Book] Madison, WI: Crop Science Society of America. Available at: https://www.crops.org/files/science-policy/cssa-crop-adaptation-position-statement.pdf (accessed on 7 July 2021).
8. Raza, A., Razzaq, A., Mehmood, S., Zou, X., Zhang, X., Lv, Y., & Xu, J., (2019). Impact of climate change on crops adaptation and strategies

to tackle its outcome: A review. *Plants,* [online] *8*(2), 34. Available at: https://www.ncbi.nlm.nih.gov/pmc/articles/PMC6409995/ (accessed on 7 July 2021).

9. Westengen, O., & Brysting, A., (2014). Crop adaptation to climate change in the semi-arid zone in Tanzania: The role of genetic resources and seed systems. *Agriculture & Food Security,* [online] *3*(1). Available at: https://agricultureandfoodsecurity.biomedcentral.com/articles/10.1186/2048-7010-3-3 (accessed on 7 July 2021).

INDEX

C

D

E

F

G